D1000258

Dynamic Heterogeneous Catalysis

Dynamic Heterogeneous Catalysis

K. TAMARU

Department of Chemistry,
University of Tokyo, Bunkyo-ku,
Tokyo, Japan

1978

ACADEMIC PRESS

London · New York · San Francisco

A SUBSIDIARY OF HARCOURT BRACE JOVANOVICH, PUBLISHERS

ACADEMIC PRESS INC. (LONDON) LTD
24–28 Oval Road,
London NW1

U.S. Edition published by
ACADEMIC PRESS INC.
111 Fifth Avenue,
New York, New York 10003

Library of Congress Catalog Card Number: 77-71841

ISBN: 0-12-684150-0

PRINTED IN GREAT BRITAIN BY
PAGE BROS (NORWICH) LTD, NORWICH

to
Sir Hugh Taylor
to whom this book owes much

Preface

Catalysis, a dominant feature in various biological reactions such as enzyme processes and photosynthesis, is vitally important in our daily life. Most of the products in chemical industry are manufactured through catalysis and, moreover, many of the difficulties which confront us today, problems of energy, natural resources and pollution are problems which are amenable to solution by catalysis. Catalysis plays an important role in pollution problems, not only in the removal of pollutants, such as NO_x, CO and sulphur compounds, but also in improving the selectivity of manufacturing processes so that undesirable by-products are not produced.

Until twenty years ago, the main approaches in the field of heterogeneous catalysis were limited to kinetic studies, isotope techniques and chemisorption measurements on each individual participating gas. It is true that the mechanism of the reaction is reflected in its kinetic behaviour, but, as discussed in this monograph, a knowledge of the kinetics alone is generally not enough to determine the reaction mechanism. Nor do chemisorption studies supply the crucial information on the mechanism of catalysis through which the overall reaction takes place. The situation, in principle, remained unchanged even after the application of i.r. spectroscopic studies was initiated twenty years ago. Chemisorption studies by themselves are, in most cases, not directly applicable to catalysis, as few adsorption measurements have been carried out during the course of reaction. Even if some of the reactants are chemisorbed on a solid surface, it does not necessarily follow that the surface will catalyse the reaction of the gas.

Nearly twenty years ago adsorption measurements during the course of catalysis were initiated by the author, and the amount, structure and reactivity of the chemisorbed species were examined under the reaction conditions along with the overall reaction rate. The study of the behaviour of chemisorbed species under the non-stationary perturbation revealed direct information concerning the mechanism of heterogeneous catalysis. Chemisorption during the reaction cannot be estimated from the adsorption measurements of each of the reactant gases measured separately over the catalyst surface. Interaction among the chemisorbed species and the location of the rate-determining step markedly influence the chemisorption and the state of the catalyst surface.

Catalysis is a branch of chemical dynamics and a catalytic reaction may be characterized as a typical chain reaction, with the catalyst behaving as the chain carrier. Any progress in the study of chemical dynamics in general may be applied to heterogeneous catalysis and vice versa. The elucidation of the mechanisms of chemical reactions through studying the behaviour of reaction intermediates should be similar, in principle, to the problem in catalysis where the behaviour of chemisorbed species is examined.

In this monograph an attempt has been made to explain the dynamic aspect of chemical reactions, heterogeneous catalysis in particular, and to emphasize the importance of the dynamic approach in gaining a deeper insight into the nature of heterogeneous catalysis. Accordingly, the monograph does not cover all aspects of heterogeneous catalysis and the author has purposely limited himself to describing experimentally proven phenomena, avoiding conjecture.

Many varieties of new tools have recently been developed for studying surface reactions, chemisorption and catalysis, and more powerful tools will be developed in the future. Accordingly, the problem of how to use these new techniques is becoming more and more important. Dynamic behaviour, such as heterogeneous catalysis, should be studied in a dynamic manner. The excitement of a football game cannot be analysed by examining the nature of the ball and each of the players. Biological systems are dynamic systems, life itself is characterized by its dynamic behaviour, and so is the entire world.

In Chapter 1, general rules for reaction rates and the mechanisms of chemical reactions are explained, heterogeneous catalysis being treated as a branch of chemical dynamics in general. Some typical modern tools for studying solid surfaces and chemisorbed species are introduced in Chapter 2 and then Chapter 3 deals with the kinetics of catalytic reactions on solid surfaces. The dynamic treatment of adsorbed species under reaction conditions using spectroscopic techniques is illustrated in Chapter 4.

I would like to express my sincere appreciation to my colleagues, particularly to Professor T. Onishi, for his collaboration in carrying out our dynamic investigations of catalysis. I am also thankful to Dr A. C. Herd of the N.Z. Fertilizer Manufacturers Research Association, New Zealand, for assistance and advice in preparing this monograph. Without his help this book would not have appeared.

Kenzi Tamaru

Kamakura
February 1977

Contents

Chapter 3
Kinetics of Catalytic Reactions on Solid Surfaces: Their Interpretation and the Elucidation of Reaction Mechanisms

Chapter 4
The Application of Spectroscopic Techniques to the Dynamic Treatment of Adsorbed Species under Reaction Conditions

Introduction

When the gases hydrogen and oxygen are mixed together at room temperature, they do not react with each other even though the mixture is thermodynamically very unstable. When nitrogen oxide is mixed with oxygen, it reacts immediately to form nitrogen dioxide, although the free energy decrease for this reaction is not as large as that for the reaction between hydrogen and oxygen.

	$\Delta G^{\circ}_{298\,K}$
$2H_2 + O_2 = 2H_2O$	$-54{\cdot}64$ kcal/mol
$2NO + O_2 = 2NO_2$	$-\ 8{\cdot}33$ kcal/mol

This example emphasizes that the thermodynamic quantities that apply to equilibrium conditions and which define "how far" a reaction will proceed should not be confused with the kinetic parameters which define "how fast" the reaction will take place.

We know a considerable amount about simple molecules, but the nature of reactions between them is very complex and far from clear. Even if molecular properties were known in much more detail, they would not explain differences in reactivities because reactivity is associated with kinetic parameters and requires dynamic measurements for its elucidation.

Similarly, "catalysis is a rate phenomenon." To elucidate the mechanisms of catalysis we must employ dynamic techniques, since, as in the case of the H_2–O_2 reaction, no true mechanism can be elucidated from static studies.

The properties of the catalytic sites on solid surfaces under working conditions are generally not the same as those of the clean surface. Under the reaction conditions some of the reactants, products, intermediates or other species are usually adsorbed on the surface, thus considerably influencing the properties of the reaction sites. The heat of adsorption and the electron accepting or donating properties (or work function) of the surface are generally affected by both the nature and amount of the adsorbed species. In some cases, such as catalysis by alloys, the composition of the catalyst at the surface is not only different from that of the bulk, but is altered by the chemisorption itself.

As the adsorbed species change during the reaction, the properties of the catalyst surface change correspondingly. For example, in the catalytic reaction between hydrogen and oxygen over a copper catalyst, it is possible that the catalytic reaction takes place by the repeated oxidation and subsequent reduction of the copper. During the reaction, the surface may be copper oxide or hydroxide or a copper surface partially covered by chemisorbed oxygen or hydroxide, any of which would be a markedly different surface from that of clean copper. In such cases, a study of the work function of a clean copper surface under an ultrahigh vacuum, for example, would have almost nothing to do with the mechanism of the catalytic reaction.

The extent of adsorption on a catalyst surface during the course of reaction cannot be estimated from separate adsorption measurements for each of the reactants and products. The nature and extent of adsorption is not only dependent upon the interaction among the surface species (such as the formation of surface complexes), but also upon the position of the rate-determining step, because this largely determines the chemical potentials of the reaction intermediates. Adsorption equilibrium is never established when the rate of removal of the reactive chemisorbed species by reaction is higher than the rate of desorption.

The decomposition of ammonia at 400°C over a transition metal catalyst, a system that will be studied in more detail in Chapter 3, provides an example of a characteristic feature of reacting systems. If nitrogen desorption is rate-determining, the chemical potential of chemisorbed nitrogen during reaction would correspond to a nitrogen pressure of 5×10^{17} atmospheres even though the ammonia and hydrogen pressures are 1×10^{-5} and 1×10^{-7} Torr respectively.

As I first pointed out in 1958, under these circumstances it should be obvious that the properties of the surface should be studied in the working state of the catalyst. Extrapolation to catalytic working conditions of any quantity derived or measured for the clean surface is usually meaningless. The amount of adsorption and the nature, structure and reactivity of the surface species under the reaction conditions have to be compared with the rate of the overall reaction. The existence of a certain adsorbed species does not guarantee that it is a reaction intermediate through which the overall reaction proceeds, neither is the most abundant adsorbed species necessarily associated with the reaction intermediate. The behaviour of each adsorbed species under the working conditions, particularly measurements of their concentrations, should be correlated with the reaction mechanism.

Such a dynamic treatment of the surface species is the only approach which leads to the elucidation of the real reaction mechanism in terms of a reaction path. The rates of each of the elementary or simple steps which make up the overall reaction path may be correlated with the properties of the catalytic sites in their working state.

The relationship between the catalytic activity and the properties of the catalytic sites and, further, the "catalyst design" may be discussed on the basis of these experimental results, rather than on the basis of observations of the input and output of a catalytic black-box or from the misleading properties of the catalyst in the absence of reactants.

Such dynamic treatments of the adsorbed species under reaction conditions may be an orthodox way to elucidate not only *how* catalytic reactions proceed, but also the reason *why* catalysis takes place.

Many new physical techniques have been developed recently: low-energy electron diffraction, photoelectron spectroscopy in the u.v. and X-ray regions, Auger electron spectroscopy, ion-neutralization spectroscopy, and Fourier-transform i.r. spectroscopy are some examples. These new techniques are providing access to extensive information about the surfaces studied, such as the arrangements of the surface atoms, elemental qualitative and quantitative analyses, energy level diagrams and the nature of the surface electron orbitals. The most efficient use of these new techniques for the elucidation of catalytic mechanisms is the next problem to be solved. The study of chemisorption by itself does not lead directly to the elucidation of the mechanisms of heterogeneous catalysis, just as the study of the properties of molecular oxygen does not lead to the elucidation of the H_2–O_2 reaction mechanism. Dynamic behaviour should be studied in a dynamic manner.

Tamaru, K. *Bull. Chem. Soc. Japan*, **31**, 666 (1958); *Adv in Catalysis*, **15**, 65 (1964).

Chapter 1

General Rules for Reaction Rates and the Mechanisms of Chemical Reactions

1-1 Overall reaction rate and the rates of the forward and backward reactions

When a (1 : 3, v/v) mixture of nitrogen and hydrogen is introduced into a closed circulation system which contains a catalyst such as the doubly promoted iron (Fe with small amounts of K_2O and Al_2O_3) at 200°C, ammonia is slowly formed.

$$3H_2 + N_2 = 2NH_3 \qquad (1\text{-}1\text{-}1)$$

The amount of ammonia formed increases with time and the rate of formation slows down as the reaction proceeds, as illustrated by the lower curve in Fig. 1. The stoichiometric mixture of nitrogen and hydrogen is not completely converted to ammonia but approaches an equilibrium value. For example when the total pressure is one atmosphere, the gas phase at equilibrium contains 15·3% NH_3.

In a similar manner, if ammonia is introduced into the system, it decomposes to hydrogen and nitrogen as shown by the upper curve in Fig. 1. At the same total pressure, the decomposition reaction gives the same equilibrium gas phase composition as the ammonia synthesis reaction. In other words, provided the final total pressure in the system is the same, reaction (1-1-1) proceeds towards the same gas phase composition regardless of whether the initial composition is pure ammonia or a stoichiometric mixture of nitrogen and hydrogen. Once the equilibrium composition is reached, the chemical composition remains unchanged as long as the temperature and pressure are kept constant.

Equilibrium compositions for the N_2–H_2–NH_3 system at one and two hundred atmospheres pressure at 200, 400 and 600°C are given in Table I.

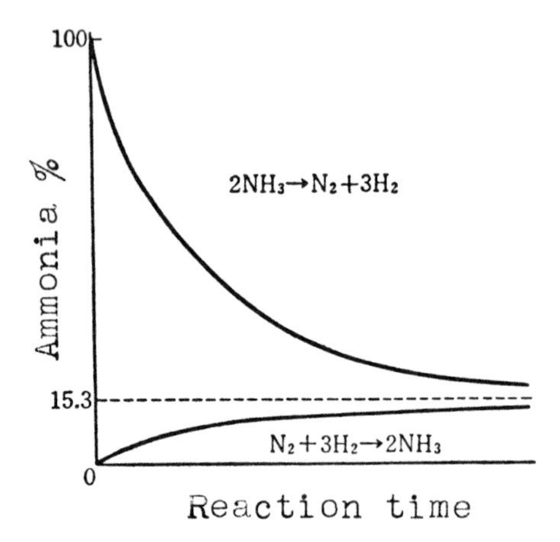

Fig. 1. Decomposition and synthesis of ammonia and their approach to an equilibrium point at 200°C under one atmosphere of pressure.

More ammonia is formed at lower temperatures and higher pressures in agreement with thermodynamic predictions for a reaction, such as ammonia synthesis, which is exothermic and which leads to a net decrease in the number of molecules.

Although the chemical composition of an equilibrium mixture of N_2–H_2–NH_3 over the catalyst does not change with time, it does not follow that no reaction is taking place. If the ^{14}N–^{14}N in the equilibrium mixture is replaced by ^{15}N–^{15}N, the dynamic nature of the equilibrium can be demonstrated. If the kinetic isotope effect is neglected, the chemical composition of the equilibrated mixture stays unchanged, but the isotope ^{15}N which was initially exclusively in the nitrogen molecules is exchanged into the ammonia molecules and correspondingly the same amount of ^{14}N is exchanged into the nitrogen molecules until the ^{15}N is completely mixed between the ammonia and the nitrogen.

Table I. The equilibrium ammonia concentrations (vol%) formed from a stoichiometric mixture of hydrogen and nitrogen at one 400 and 600°C under equilibrium total pressures of one and two hundred atmospheres

Temperature	Pressure	
	1 atm	200 atm
200°C	15·3%	85·8%
400	0·44	36·3
600	0·049	8·25

The isotope mixing in the equilibrated mixture clearly demonstrates that ammonia synthesis (forward reaction), from $^{15}N{-}^{15}N$ and hydrogen to form $^{15}NH_3$, and ammonia decomposition (backward reaction) to form nitrogen and hydrogen, are both taking place. Since the chemical composition does not change, the rates of the two reactions must be equal. In other words, the rates of the forward (V_+) and backward (V_-) reactions are equal at equilibrium and are given by the rate of the isotope mixing.

When mixing is complete, all of the following processes are at equilibrium and the system will have the composition given in Table II.

$$^{15}N{-}^{15}N + 3H_2 = 2\,^{15}NH_3$$

$$^{15}NH_3 + {}^{14}NH_3 = {}^{15}N{-}^{14}N + 3H_2$$

$$2\,^{14}NH_3 = {}^{14}N{-}^{14}N + 3H_2$$

Since the equilibrium composition (1 atm, 200°C) is 21·2 % N_2, 63·5 % H_2, and 15·3 % NH_3, the fraction of nitrogen atoms that are ^{15}N is 21·2 × 2/(21·2 × 2 + 15·3). When the isotope mixing has reached equilibrium, the probability that a nitrogen atom is ^{15}N is independent of the molecular species (nitrogen or ammonia) in which it is found. Accordingly, the ratio of $^{14}NH_3$ to $^{15}NH_3$ in the final composition is 15·3:21·2 × 2, and the ratio†

Table II. The variation in the isotopic ratio in nitrogen and ammonia with time[a]

	NH_3(15·3 %)		N_2(21·2%)		
	$^{14}NH_3$	$^{15}NH_3$	$^{15}N{-}^{15}N$	$^{15}N{-}^{14}N$	$^{14}N{-}^{14}N$
$t = 0$	100%	0%	100%	0%	0%
$t = \infty$	$100X$	$100Y$	$100X^2$	$200XY$	$100Y^2$

[a] $X = 15{\cdot}3/(21{\cdot}2 \times 2 + 15{\cdot}3)$; $Y = 21{\cdot}2 \times 2/(21{\cdot}2 \times 2 + 15{\cdot}3)$.

$$^{15}N{-}^{15}N : {}^{15}N{-}^{14}N : {}^{14}N{-}^{14}N = (21{\cdot}2 \times 2)^2 : 2 \times 2 \times 15{\cdot}3 \times 21{\cdot}2 : (15{\cdot}3)^2$$

In general, the rate of change in the overall composition (molecules/time) is

† When we pick up two balls from a box which contains n red balls and m white balls uniformly mixed, the probability that the two balls are both red is $(n/(n + m))^2$, that they are both white is $(m/(n + m))^2$, and that they are red and white is $2nm/(n + m)^2$.

given by the difference between the rates of the forward and backward reactions. At equilibrium, the overall rate is zero because V_+ is equal to V_-.

$$V = V_+ - V_- \tag{1-1-2}$$

In the reacting system N_2–H_2–NH_3 containing ^{15}N and ^{14}N, the rate of $^{15}NH_3$ formation is given by the equation (1-1-3), where Z^A is the fraction of ^{15}N in the ammonia, Z^N the fraction of ^{15}N in the nitrogen and **a** is the number of ammonia molecules in the reacting system.

$$\mathrm{d}(^{15}NH_3)/\mathrm{d}t = \mathrm{d}(\mathbf{a}Z^A)/\mathrm{d}t = 2(V_+Z^N - V_-Z^A) \tag{1-1-3}$$

V_+Z^N represents the rate of $^{15}NH_3$ formation from nitrogen molecules containing ^{15}N, while V_-Z^A represents the rate of $^{15}NH_3$ decomposition. The coefficient two on the right-hand side of equation (1-1-3) is included because V_+ and V_- represent the number of times reaction (1-1-1) takes place per unit time, and each time reaction (1-1-1) occurs, two molecules of ammonia are formed.

Of course, equation (1-1-3) is also valid for a non-equilibrated reaction system. V is the rate of reaction (1-1-1) and corresponds to half of the rate of ammonia formation $\mathrm{d}\mathbf{a}/\mathrm{d}t$:

$$V = V_+ - V_- = (\tfrac{1}{2})(\mathrm{d}\mathbf{a}/\mathrm{d}t) \tag{1-1-2'}$$

Equations (1-1-2′) and (1-1-3) give the following equation:

$$\begin{aligned}\mathrm{d}(\mathbf{a}Z^A)/\mathrm{d}t &= \mathbf{a}(\mathrm{d}Z^A/\mathrm{d}t) + Z^A(\mathrm{d}\mathbf{a}/\mathrm{d}t)\\ &= 2[V_+Z^N - (V_+ - (\tfrac{1}{2})(\mathrm{d}\mathbf{a}/\mathrm{d}t))Z^A]\\ &= 2V_+Z^N - 2V_+Z^A + Z^A(\mathrm{d}\mathbf{a}/\mathrm{d}t)\end{aligned}$$

Consequently, by removing $Z^A(\mathrm{d}\mathbf{a}/\mathrm{d}t)$ from both sides,

$$\mathbf{a}(\mathrm{d}Z^A/\mathrm{d}t) = 2V_+(Z^N - Z^A)$$

$$V_+ = \frac{\mathbf{a}(\mathrm{d}Z^A/\mathrm{d}t)}{2(Z^N - Z^A)} \tag{1-1-4}$$

From equation (1-1-2′),

$$V_+(1 - (V_-/V_+)) = (\tfrac{1}{2})(\mathrm{d}\mathbf{a}/\mathrm{d}t)$$

Rearranging and substituting V_+ from equation (1-1-4),

$$V_+/V_- = \frac{V_+}{V_+ - (\frac{1}{2})(\mathbf{da}/dt)}$$

$$= \frac{1}{1 - (1/\mathbf{a})(\mathbf{da}/dZ^A)(Z^N - Z^A)}$$

$$= \frac{1}{1 - (d \ln P_A/dZ^A)(Z^N - Z^A)} \qquad (1\text{-}1\text{-}5)$$

This shows that at time infinity, Z^N becomes equal to Z^A and the ratio of V_+ and V_- reaches unity. The ammonia pressure, P_A, Z^A and Z^N in equation (1-1-5) may be continuously observed during the course of the reaction. V can also be measured by determining the number of ammonia molecules, **a**, in the reacting system. Consequently, the difference between V_+ and V_- and the ratio of V_+ and V_-, equation (1-1-5), can be experimentally determined to give V, V_+ and V_- during the course of the reaction under various experimental conditions.[1]

1-2 Rate equations

The rate of the forward reaction, V_+ is in general a function of the temperature T and the pressure P_i of species i which participates in the reaction:

$$V_+ = f(T, P_i, \ldots) \qquad (1\text{-}2\text{-}1)$$

In many cases the rate can be expressed in the form of the law of mass action:

$$V_+ = k\prod_i P_i^{\alpha_i} \qquad (1\text{-}2\text{-}2)\dagger$$

where k, the rate constant, is a function of temperature but not of P_i and α_i is the order of reaction with respect to component i.

† The rate of a catalytic reaction based on the Langmuir model, as will be shown later, involves an expression involving algebraic sum of terms in the numerator and/or denominator. But in some cases it may be approximated, for example, as follows, $\frac{bp}{1 + bp} \approx (bp)^n$, in a limited pressure range.

The dependence of k upon T is generally expressed by the "Arrhenius expression":

$$k = A\exp(-E/RT) \tag{1-2-3}$$

where A is called the pre-exponential factor and E the apparent activation energy.

In heterogeneous catalysis, it has been shown that for some series of related systems, (either a single reaction on a series of catalysts, or a series of reactions on a single catalyst) A and E vary in the same direction, thus partially compensating each other to give similar activities. This is frequently called the Theta Rule or compensation effect and may be expressed as follows:

$$\ln A = \text{const.} + E/R\theta \tag{1-2-4}$$

where θ is the temperature at which all k are identical. This is easily understood by substituting A from equation (1-2-4) into equation (1-2-3).

The order of reaction with respect to each of the components is generally not predictable from the stoichiometric equation of the chemical reaction. Some typical examples of the decomposition of hydrogen compounds are given in Table III.

Table III. Rate equations for the decomposition of hydrogen compounds over various catalysts

Reaction	Catalyst	Rate equation
$2HI \rightarrow H_2 + I_2$	none	$V = k(HI)^2$
	Pt	$V = k'(HI)$
	Au	$V = k''$
$AsH_3 \rightarrow As + (\frac{3}{2})H_2$	As	$V = k(AsH_3)$
$SnH_4 \rightarrow Sn + 2H_2$	Sn	$V = k(SnH_4)$
$SbH_3 \rightarrow Sb + (\frac{3}{2})H_2$	Sb	$V = k(SbH_3)^n (1{\cdot}0 \geqslant n > 0{\cdot}6)$
$GeH_4 \rightarrow Ge + 2H_2$	Ge	$V = k$

Another example is the decomposition of ammonia over metal catalysts. The rate of ammonia decomposition over an ammonia synthesis catalyst is expressed by the following equation:

$$-\mathrm{d}(NH_3)/\mathrm{d}t = kP_{NH_3}^{0{\cdot}6}/P_{H_2}^{0{\cdot}85} \tag{1-2-5}$$

The decomposition rate over an iron catalyst is expressed:

$$-\mathrm{d}(\mathrm{NH_3})/\mathrm{d}t = kP_{\mathrm{NH_3}}/P_{\mathrm{H_2}}^{1\cdot5} \tag{1-2-6}$$

These equations are very different from the ones that might be expected from reaction (1-1-1).

Generally speaking, the rate equation for V_+ in the chemical reaction

$$aA + bB = mM + nN \tag{1-2-7}$$

may be expressed as follows; as in equation (1-2-2):

$$V_+ = k_+ P_A^{\alpha} P_B^{\beta} P_M^{\lambda} P_N^{\delta} \tag{1-2-8}$$

If the backward rate V_- is given by the following equation,

$$V_- = k_- P_A^{w} P_B^{x} P_M^{y} P_N^{z} \tag{1-2-9}$$

are those coefficients, w, x, y and z to be uniquely determined?†

† It should be noted here that the correlation between kinetics and thermodynamics is frequently explained as follows, for example, by employing the decomposition and formation of hydrogen iodide as a model reaction:

$$2\mathrm{HI} = \mathrm{H_2} + \mathrm{I_2}$$

$$V_+ = k_+(\mathrm{HI})^2$$

$$V_- = k_-(\mathrm{H_2})(\mathrm{I_2})$$

At equilibrium,

$$V_+ = V_-$$

Consequently,

$$k_+(\mathrm{HI})^2 = k_-(\mathrm{H_2})(\mathrm{I_2})$$

or

$$\frac{(\mathrm{H_2})(\mathrm{I_2})}{(\mathrm{HI})^2} = \frac{k_+}{k_-}$$

which is equal to the equilibrium constant, K. Such straightforward treatment cannot be a general one, which is easily understood if we think of the reaction between hydrogen and bromine as will be explained later.

Suppose the rate equations for V_+ and V_- are valid and are independent of the distance from equilibrium. At equilibrium, V_+ is equal to V_- and consequently,

$$k_+/k_- = P_A^{w-\alpha} P_B^{x-\beta} P_M^{y-\gamma} P_N^{z-\delta} \tag{1-2-10}$$

On the other hand, the equilibrium constant K is defined by:

$$K = P_A^{-a} P_B^{-b} P_M^{m} P_N^{n} \tag{1-2-11}$$

Accordingly,

$$\frac{w-\alpha}{-a} = \frac{x-\beta}{-b} = \frac{y-\gamma}{m} = \frac{z-\delta}{n} \tag{1-2-12}$$

Thus, the experimental measurement of α, β, γ and δ is not enough to determine the coefficients w, x, y and z. However, if one of w, x, y and z is known, then they are all determined. In the case of ammonia decomposition over the synthesis catalyst, V_+ is given by equation (1-2-5) and provided V_- is proportional to P_{N_2}, the backward rate equation is:

$$V_- = k_- P_{N_2} P_{H_2}^{2\cdot15} / P_{NH_3}^{1\cdot4} \tag{1-2-13}$$

1-3 Overall reaction, elementary reaction and rate-determining step

Suppose we have a consecutive reaction†

$$A \to B \to C \tag{1-3-1}$$

where A is converted into C through a reaction intermediate B. If the rates of the two steps are proportional to the concentration of the reactants and the rate constants for A→B and B→C are k_1 and k_2, respectively, the following equations are obtained:

$$\mathrm{d(A)}/\mathrm{d}t = -k_1\mathrm{(A)}$$

$$\mathrm{d(B)}/\mathrm{d}t = k_1\mathrm{(A)} - k_2\mathrm{(B)} \tag{1-3-2}$$

$$\mathrm{d(C)}/\mathrm{d}t = k_2\mathrm{(B)}$$

† A reaction which consists of more than two steps is called a complex reaction, while that which cannot be divided into simpler processes is called an elementary step.

The rate of B formation is the difference between the rate that B is supplied from A and the rate at which B is consumed to form C. Equations (1-3-2) can be integrated to obtain the dependences of (A), (B) and (C) upon the reaction time as follows:

$$(\mathrm{A}) = (\mathrm{A})_0\, \mathrm{e}^{-k_1 t}$$

$$(\mathrm{B}) = \frac{k_1(\mathrm{A})_0}{k_2 - k_1}(\mathrm{e}^{-k_1 t} - \mathrm{e}^{-k_2 t}) \qquad (1\text{-}3\text{-}3)$$

$$(\mathrm{C}) = (\mathrm{A})_0(1 - \mathrm{e}^{-k_1 t}) - \frac{k_1(\mathrm{A})_0}{k_2 - k_1}(\mathrm{e}^{-k_1 t} - \mathrm{e}^{-k_2 t})$$

where $(\mathrm{A})_0$ is the concentration of A at time zero and it has been assumed that (B) and (C) are zero at time zero and also that $k_2 \neq k_1$. Fig. 2 schematically shows the concentration changes with time. Throughout the reaction, the total number of molecules in the reacting system remains unchanged:

$$(\mathrm{A}) + (\mathrm{B}) + (\mathrm{C}) = (\mathrm{A})_0 \qquad (1\text{-}3\text{-}4)$$

Let us examine the change of (B) with time. According to equation (1-3-2), if $k_1(\mathrm{A}) > k_2(\mathrm{B})$, then $\mathrm{d(B)}/\mathrm{d}t > 0$; if $k_1(\mathrm{A}) = k_2(\mathrm{B})$, $\mathrm{d(B)}/\mathrm{d}t = 0$; and if $k_1(\mathrm{A}) < k_2(\mathrm{B})$, $\mathrm{d(B)}/\mathrm{d}t < 0$. In the initial stage of the reaction, as $k_1(\mathrm{A}) > k_2(\mathrm{B})$,

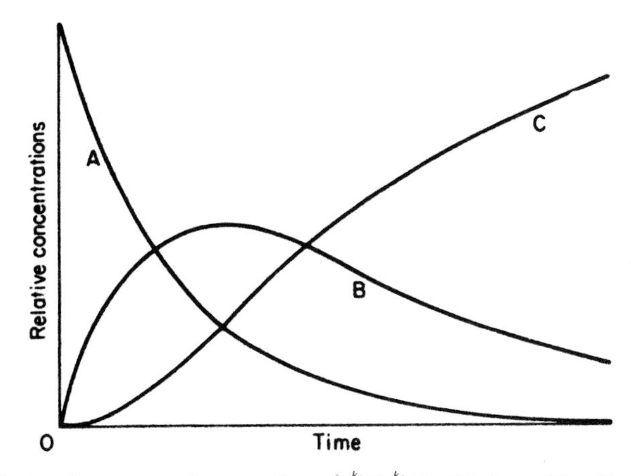

Fig. 2. Two first-order consecutive reactions $\mathrm{A}\xrightarrow{k_1}\mathrm{B}\xrightarrow{k_2}\mathrm{C}$ with $k_1 = 2k_2$. Note the zero slope of curve C at zero time, the maximum in curve B and the inflection point in curve C.

(B) increases with time, while (A) decreases until (B) reaches a maximum at $k_1(A) = k_2(B)$. Thereafter (B) decreases toward zero if the reaction ends at (A) = (B) = 0. The time at which (B) reaches a maximum, t_m, may be obtained as follows,

$$\begin{aligned} k_1(A) &= k_1(A)_0\, e^{-k_1 t_m} \\ &= k_2(B) \\ &= \frac{k_1 k_2 (A)_0}{k_2 - k_1}(e^{-k_1 t_m} - e^{-k_2 t_m}) \end{aligned}$$

Consequently,

$$t_m = \frac{1}{k_2 - k_1} \ln(k_2/k_1) \tag{1-3-5}$$

and (B) at t_m, $(B)_m$, is given by:

$$(B)_m/(A)_0 = (k_1/k_2)^{k_2/(k_2 - k_1)} \tag{1-3-6}$$

Accordingly, the ratio $(B)_m/(A)_0$ gets smaller as k_1/k_2 becomes smaller. If the reaction intermediate, B, is much more reactive than A, $k_2 \gg k_1$. In consequence, $(B)_m$ becomes very small and the curve for (B) in Fig. 2 becomes flat, almost coinciding with the abscissa. In other words, under these circumstances $d(B)/dt \doteqdot 0$. If we put $k_2 \gg k_1$ into equation (1-3-3), we obtain the following equations,

$$\begin{aligned} (B) &= (k_1/k_2)(A)_0 e^{-k_1 t} \\ (C) &= (A)_0(1 - e^{-k_1 t}) \end{aligned} \tag{1-3-7}$$

In such a situation, where (B) is much smaller than (A) and (C), (B) can be neglected throughout the reaction and consequently:

$$(A)_0 = (A) + (C): \; -d(A)/dt \doteqdot d(C)/dt \tag{1-3-8}$$

From equations (1-3-2) and (1-3-8)

$$k_1(A) = k_2(B)$$

$$(B) = (k_1/k_2)(A)_0\, e^{-k_1 t} \tag{1-3-7'}$$

$$(C) = (A)_0(1 - e^{-k_1 t})$$

These results are in agreement with equation (1-3-7). Such a treatment, which assumes that for B, the reactive reaction intermediate, $d(B)/dt \doteq 0$ is sometimes called the "steady-state approximation" and such a situation is called the "steady-state". Under the conditions of steady-state, the rate of creation of each intermediate may be considered to be balanced with that of its consumption in the course of the overall reaction.

In the case where $k_1 = k_2$, the following equations may be obtained in a similar manner.

$$(A) = (A)_0 e^{-k_1 t}$$

$$(B) = k_1 (A)_0 t\, e^{-k_1 t}$$

$$(B)_m = (A)_0 / e \qquad (1\text{-}3\text{-}9)$$

$$t_m = 1/k_1$$

$$(C) = (A)_0 (1 - e^{-k_1 t}(1 + k_1 t))$$

If the steps of the consecutive reaction are not unidirectional as in (1-3-1), but are reversible as in the following reaction, the kinetic behaviour can be treated in a similar fashion. If B is much more reactive than A or C, as soon as B is formed by either step it changes rapidly to A or C. In consequence, (B) is always negligibly small

$$A \overset{1}{\rightleftarrows} B \overset{2}{\rightleftarrows} C \qquad (1\text{-}3\text{-}10)$$

in comparison with (A) or (C), and $d(B)/dt$ may be considered to be zero as in a steady-state approximation.

In such a case,

$$V = v_{+1} - v_{-1} = v_{+2} - v_{-2}$$

$$= k_{+1}(A) - k_{-1}(B) = k_{+2}(B) - k_{-2}(C) \qquad (1\text{-}3\text{-}11)$$

where v_{+1}, v_{-1} and v_{+2}, v_{-2} represent the forward and backward reaction rates for the first and the second steps respectively, and the corresponding rate constants are written as k_{+1}, k_{-1} and k_{+2}, k_{-2}, as shown in equation (1-3-11).

From equation (1-3-11),

$$(B) = \frac{k_{+1}(A) + k_{-2}(C)}{k_{-1} + k_{+2}} \qquad (1\text{-}3\text{-}12)$$

Substituting equation (1-3-12) into (1-3-11),

$$V = \frac{1}{k_{-1} + k_{+2}}\left(k_{+1}k_{+2}(\mathrm{A}) - k_{-1}k_{-2}(\mathrm{C})\right)$$

$$= V_{+\mathrm{S}} - V_{-\mathrm{S}} \tag{1-3-13}$$

where $V_{+\mathrm{S}}$ is the unidirectional rate at which the reaction proceeds from A via B to C without being changed back to A again. Similarly, $V_{-\mathrm{S}}$ is the rate of the backward unidirectional reaction from C to A.†

(i) if $k_{-1} \gg k_{+2}$,

$$V_{+S} = (k_{+1}/k_{-1})k_{+2}(\mathrm{A})$$
$$V_{-S} = k_{-2}(\mathrm{C}) \tag{1-3-14}$$

In this case, $v_{+2}/(v_{-1} + v_{+2}) \ll 1$, and most of the B that is formed from A is reconverted back to A again without being further converted to C. The ratio v_{+1}/v_{-1} approaches to unity when $k_{-1} \gg k_{+2}$:

$$\frac{v_{+1}}{v_{-1}} = \frac{k_{+1}(\mathrm{A})}{k_{-1}(\mathrm{B})} = \frac{(k_{-1} + k_{+2})k_{+1}(\mathrm{A})}{\left(k_{+1}(\mathrm{A}) + k_{-2}(\mathrm{C})\right)k_{-1}} \doteqdot 1 \tag{1-3-15}$$

Accordingly, if $k_{-1} \gg k_{+2}$, then $v_{-1} \gg v_{+2}$ and $v_{+1}/v_{-1} \doteqdot 1$. In other words, the first step takes place in both directions frequently, while the second step takes place comparatively rarely. These relative rates are schematically represented in Fig. 3, where v_{+1} and v_{-1} are both much faster than v_{+2} and v_{-2}, although the differences, $v_{+1} - v_{-1}$ and $v_{+2} - v_{-2}$ are the same.

† If the overall reaction takes place via a single route with many elementary steps, 1, 2, 3, . . . , i, . . . , n, the ratio of the forward unidirectional rate of the single route reaction (V_{+S}) to the backward one (V_{-S}) is given in general by

$$\frac{V_{+S}}{V_{-S}} = \frac{v_{+1}\left(\frac{v_{+2}}{v_{-1} + v_{+2}}\right)\left(\frac{v_{+3}}{v_{-2} + v_{+3}}\right)\cdots\left(\frac{v_{+i}}{v_{-(i-1)} + v_{+i}}\right)\cdots\left(\frac{v_{+n}}{v_{-(n-1)} + v_{+n}}\right)}{v_{-n}\left(\frac{v_{-(n-1)}}{v_{+n} + v_{-(n-1)}}\right)\cdots\left(\frac{v_{-i}}{v_{+(i+1)} + v_{-i}}\right)\cdots\left(\frac{v_{-1}}{v_{+2} + v_{-1}}\right)}$$

$$= \frac{v_{+1}v_{+2}v_{+3}\ldots}{v_{-1}v_{-2}v_{-3}} \frac{v_{+1}}{v_{-1}} \cdots \frac{v_{+n}}{v_{-n}}$$

$$= \prod (v_{+i}/v_{-i})$$

Fig. 3. A consecutive reaction, $A \overset{1}{\rightleftarrows} B \overset{2}{\rightleftarrows} C$ in a steady-state,

$$V = v_{+1} - v_{-1} = v_{+2} - v_{-2}$$

where v_{+1}, v_{-1} and v_{+2}, v_{-2} represent the forward and backward reaction rates of the first and second steps, respectively.

Generally speaking, when an elementary reaction (1-3-16) takes place, the free energy drop, ΔG, which accompanies the reaction is related to the ratio of the forward and backward rates.

$$A = B \tag{1-3-16}$$

At equilibrium, $v_+ = v_- = k_+(A)_e = k_-(B)_e$, where the subscript e represents the equilibrium condition. The equilibrium constant, K, is equal to k_+/k_- which equals $(B)_e/(A)_e$. K is also given by $\exp(-\Delta G_0/RT)$, where ΔG_0 is the standard free energy increase in the reaction under the standard conditions, $(A) = (B) = 1$.

$$K = k_+/k_- = (B)_e/(A)_e = \exp(-\Delta G_0/RT) \tag{1-3-17}$$

Under the reaction conditions:

$$-\Delta G = -\Delta G_0 + RT\ln\frac{(A)}{(B)}$$

where $-\Delta G$ corresponds to the driving force of the reaction. Accordingly,

$$-\Delta G = -\Delta G_0 + RT\ln\left(\frac{k_+(A)}{k_-(B)}\frac{1}{K}\right)$$

$$= -\Delta G_0 + RT\ln\frac{v_+}{v_-} - RT\ln K$$

$$= RT\ln(v_+/v_-) \tag{1-3-18}$$

This is the relation between the driving force of the reaction (free energy drop) and v_+/v_-. If v_+/v_- is unity, ΔG becomes zero, in other words, the elementary reaction system is at equilibrium.

Let us return to reaction (1-3-10) which is in the condition represented by Fig. 3. In this case, as $v_{+1}/v_{-1} \doteqdot 1$, the first step of the reaction is in equilibrium and no free energy drop is involved even though the overall reaction is proceeding. Accordingly, the total free energy drop which accompanies the overall reaction, A→C, is mainly associated with the second step.

To explain the situation in a different manner, let us consider a series of three water tanks connected by means of two different sized tubes, (1) and (2), as depicted in Fig. 4. Tank A on the left corresponds to the initial reactant system

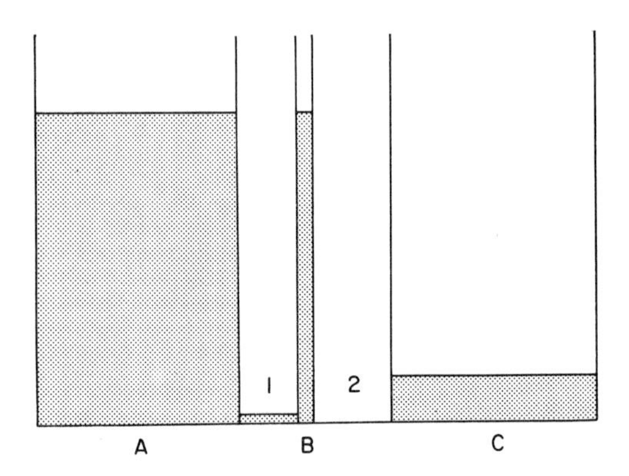

Fig. 4. A series of three water tanks connected by means of two different sized tubes.

A, while tank C on the right corresponds to the reaction product system. The water level of tank A is higher than that of C and water flows from left to right (analogous to the reaction proceeding). The small tank B in the middle represents the reaction intermediate. The capacities of the individual tanks correspond to the residence times. In this model, the water level of each tank corresponds to the chemical potential of each system. If the water levels of adjacent tanks are equal, they are in equilibrium.

If tube (2) is much narrower than tube (1), the water levels of the tanks will adjust as shown in Fig. 4. This is the case for the reacting system under the conditions shown in Fig. 3. The difference between the water levels at the narrower tube is equal to that between A and C, and this is the water level (free energy) difference causing the overall flow (reaction). The step where the free energy drop takes place is called the "rate-determining step" and the location of the difference in the water level can accordingly be used as a criterion of the rate-determining step. Consequently, one method of identifying the rate-determining step is to determine the water level (chemical potential) of each tank during the course of the overall reaction. In Figs 3 and 4, the second step is the rate-determining one provided $k_{-1} \gg k_{+2}$.

By analogy, if k_{-1} is much smaller than k_{+2}, the second step attains equilibrium during the course of the reaction and the first step is the rate-determining step. The water levels of tanks B and C become equal, i.e. reach equilibrium, while the overall reaction is taking place.

The water level of the reaction intermediate, B, depends markedly on the location of the rate-determining step. If the first step is rate-determining, the level of B is equal to the level of tank C, whereas if the second step is rate-determining, it is equal to that of tank A.

If the overall reaction is composed of a series of more than two elementary steps,

$$L \overset{1}{\rightleftharpoons} I_1 \overset{2}{\rightleftharpoons} I_2 \overset{3}{\rightleftharpoons} \cdots \overset{n-1}{\rightleftharpoons} I_{n-1} \overset{n}{\rightleftharpoons} R \tag{1-3-19}$$

the tank model may be treated in a similar fashion as shown in Fig. 5. If one of the steps is much slower than all of the others, all the water levels preceding

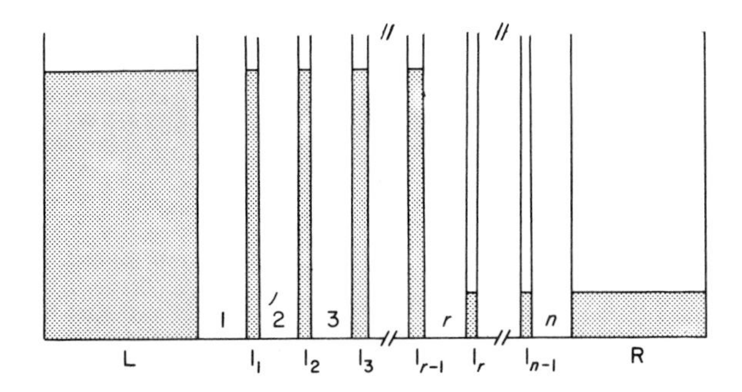

Fig. 5. A series of $n + 1$ water tanks connected by tubes of various sizes. (The rth tube is the narrowest tube.)

the rate-determining step (the narrowest tube) are equal and the same goes for the levels following the rate-determining step. The difference between the water levels at the narrowest tube is equal to the difference in levels between the reactants and the final products.

In the steady state, the rate of the overall reaction is the difference between the forward and backward rates of each step divided by its "stoichiometric number" (s). This is the number of times that the step must be repeated in order to obtain, by summation over all the steps, the overall stoichiometric equation for the reaction as written:†

$$V = \frac{1}{s_1}(v_{+1} - v_{-1}) = \frac{1}{s_2}(v_{+2} - v_{-2}) = \cdots$$

$$= \frac{1}{s_i}(v_{+i} - v_{-i}) = \cdots = \frac{1}{s_n}(v_{+n} - v_{-n}) \qquad (1\text{-}3\text{-}20)$$

From equations (1-3-17) and (1-3-20),

$$V = (v_{+1}/s_1)(1 - \exp(\Delta G_1/RT)) = \cdots = \frac{(1 - \exp(\Delta G_1/RT))}{s_1/v_{+1}}$$

$$= \cdots = \frac{(1 - \exp(\Delta G_i/RT))}{s_i/v_{+i}} = \cdots = \frac{(1 - \exp(\Delta G_n/RT))}{s_n/v_{+n}} \qquad (1\text{-}3\text{-}23)$$

where $\Delta G_1, \Delta G_2, \ldots, \Delta G_n$ are the free energy differences for each of the steps. According to equation (1-3-23), if v_{+i}/s_i is large enough, ΔG_i approaches zero, which means that the ith step is practically at equilibrium.

† If the reaction $2B_2 + A_2 = 2AB_2$ proceeds via the following two steps, the stoichiometric number of the first step is unity, while that of the second is two:

		s
(1)	$A_2 = 2A$	1
(2)	$A + B_2 = AB_2$	2
	$2B_2 + A_2 = 2AB_2$	

(1-3-21)

In the steady-state,

$$V = v_{+1} - v_{-1} = \tfrac{1}{2}(v_{+2} - v_{-2}) \qquad (1\text{-}3\text{-}20')$$

v_1 and v_2 are related to the number of times that steps (1) and (2) take place respectively; the second step must proceed twice for the overall reaction to occur once.

If ammonia synthesis, $N_2 + 3H_2 = 2NH_3$, took place via the following steps,

		s
(1)	$N_2(g) \rightarrow 2N(a)$	1
(2)	$H_2(g) \rightarrow 2H(a)$	3
(3)	$N(a) + H(a) \rightarrow NH(a)$	2
(4)	$NH(a) + H_2(g) \rightarrow NH_3(g)$	2

(1-3-22)

the stoichiometric numbers of the individual steps would be as given above.

The free energy change for the overall reaction, ΔG, is given by the following equation:

$$\Delta G = \sum_{1}^{n} s_i \Delta G_i \qquad (1\text{-}3\text{-}24)$$

If the rate of one of the steps (rth step) is markedly slower than any of the others,

$$s_r/v_{+r} \gg s_j/v_{+j} \qquad (j \neq r) \qquad j = 1, 2, \ldots, r-1, r+1, \ldots, n \qquad (1\text{-}3\text{-}25)$$

all of the steps except the rth step should be practically at equilibrium, in which case,

$$V = v_{+r}(1 - \exp(\Delta G/s_r RT)) \qquad (1\text{-}3\text{-}26)$$

or

$$v_{+r}/v_{-r} = \exp(-\Delta G/s_r RT)$$

and

$$-\Delta G = -s_r \Delta G_r \quad \text{or} \quad \Delta G_r = \Delta G/s_r \qquad (1\text{-}3\text{-}27)$$

Accordingly,

$$V = \frac{1}{s_r}(v_{+r} - v_{-r}) = \frac{v_{+r}}{s_r}(1 - \exp(\Delta G/s_r RT)) \qquad (1\text{-}3\text{-}28)$$

In this case, if the concentrations of the reactants and products are unity, the rates become equal to the rate constants and the free energy difference is what is known as the standard free energy, ΔG_0; accordingly,

$$V_+/V_- = k_+/k_- = v_{+r}/v_{-r} = \exp(-\Delta G_0/s_r RT) = K^{(1/s_r)}$$

or

$$k_+/k_- = K^{(1/s_r)} \qquad (1\text{-}3\text{-}29)\dagger$$

The ratio of the forward and backward rate constants is equal to the equilibrium constant to the power of the reciprocal of the stoichiometric number of the rate-determining step.

† This equation was first proposed in an explicit manner by Horiuti.[1]

In the case of reaction (1-3-21), if step (1) is rate-determining, step (2) is in equilibrium:

$$k_{+2}(A)(B_2) = k_{-2}(AB_2)$$

or

$$(A) = \frac{k_{-2}}{k_{+2}} \frac{(AB_2)}{(B_2)}$$

Consequently, the rates of forward and backward reactions, V_+ (in terms of the number of AB_2 molecules formed per unit time) and V_-, respectively, are expressed as follows:

$$V_+ = 2k_{+1}(A_2) = k_+(A_2)$$

$$V_- = 2k_{-1}(A)^2 = 2k_{-1}\left(\frac{k_{-2}}{k_{+2}}\right)^2 \frac{(AB_2)^2}{(B_2)^2} = k_- \frac{(AB_2)^2}{(B_2)^2}$$

where k_+ and k_- are the rate constants of forward and backward reaction, respectively, and the coefficient two is included because two AB_2 molecules are formed if step (1) takes place once.

If step (2) is rate-determining, step (1) is in equilibrium:

$$(A) = \left[\frac{k_{+1}}{k_{-1}}(A_2)\right]^{\frac{1}{2}}$$

and

$$V_+ = k_{+2}(A)(B_2) = k_{+2}\left(\frac{k_{+1}}{k_{-1}}\right)^{\frac{1}{2}} (A_2)^{\frac{1}{2}} (B_2) = k_+(A_2)^{\frac{1}{2}} (B_2)$$

$$V_- = k_{-2}(AB_2) = k_-(AB_2)$$

Accordingly, the kinetic expression depends upon which step is rate-determining.

When step (1) is rate-determining, the ratio of the rate constants (k_+/k_-) for the forward and backward reactions is

$$\frac{k_+}{k_-} = \frac{k_{+1}}{k_{-1}}\left(\frac{k_{-2}}{k_{+2}}\right)^2 = \frac{(AB_2)^2}{(B_2)^2(A_2)} = K$$

However, when step (2) is rate-determining, it is

$$\frac{k_+}{k_-} = \frac{k_{+2}}{k_{-2}}\left(\frac{k_{+1}}{k_{-1}}\right)^{\frac{1}{2}} = \frac{(AB_2)}{(B_2)(A_2)^{\frac{1}{2}}} = K^{\frac{1}{2}}$$

which is in accord with equation (1-3-29) as the stoichiometric number of the second step is two.

If we write the overall reaction as

$$B_2 + A_2 = AB_2$$

the equilibrium constant (K') is expressed as follows:

$$K' = \frac{(AB_2)}{(B_2)(A_2)^{\frac{1}{2}}}$$

and the stoichiometric number of the first step is half, while that of the second is unity. Accordingly, when step (1) is rate-determining,

$$\frac{k_+}{k_-} = \frac{k_{+1}}{k_{-1}\left(\dfrac{k_{-2}}{k_{+2}}\right)^2} = K'^2 = \frac{(AB_2)^2}{(B_2)^2(A_2)}$$

If step (2) is rate-determining,

$$\frac{k_+}{k_-} = \frac{k_{+2}}{k_{-2}}\left(\frac{k_{+1}}{k_{-1}}\right)^{\frac{1}{2}} = K'$$

In all cases, accordingly, it is demonstrated that equation (1-3-29) is valid.

As the apparent activation energy is given by the following equation;

$$\mathrm{d}\ln k/\mathrm{d}T = E/RT^2 \tag{1-3-30}$$

the difference between the apparent activation energies for the forward and backward reactions is as follows:

$$E_+ - E_- = RT^2\,\mathrm{d}\ln k_+/\mathrm{d}T - RT^2\mathrm{d}\ln k_-/\mathrm{d}T = \frac{1}{s_r}RT^2\mathrm{d}\ln K/\mathrm{d}T \tag{1-3-31}$$

On the other hand, from thermodynamics, we know that

$$RT^2\mathrm{d}\ln K/\mathrm{d}T = \Delta H_{\mathrm{RL}} \tag{1-3-32}$$

where ΔH_{RL} is the standard enthalpy increase in the overall reaction. Thus, the difference in the activation energies for the forward and backward reactions is equal to $\Delta H_{RL}/s_r$. From equation (1-3-26)

$$s_r = \frac{-\Delta G}{RT\ln(v_{+r}/v_{-r})} = \frac{-\Delta G}{RT\ln(V_+/V_-)} \tag{1-3-33}$$

This is the equation which enables us to determine the stoichiometric number of the rate-determining step. The stoichiometric number thus determined may give, in some cases, crucial information as to the location of the rate-determining step. For example, for ammonia synthesis as represented by the steps given by equation (1-3-22), the stoichiometric numbers for the individual steps range from one to three, and by measuring the value of s_r, we can differentiate between step (1) and the other three steps in (1-3-22). In this case,

$$-\Delta G = RT\ln\left(\frac{P_{NH_3}^2}{P_{N_2}P_{H_2}^3}\frac{1}{K}\right) \tag{1-3-34}$$

From equations (1-3-33) and (1-1-6),

$$s_r = \frac{\log\left(\dfrac{P_{NH_3}^2}{P_{N_2}P_{H_2}^3}\dfrac{1}{K}\right)}{\log\left(1 - \dfrac{d\ln P_{NH_3}}{dZ^A}(Z^N - Z^A)\right)} \tag{1-3-35}$$

Every term in the right-hand side of equation (1-3-35) may be experimentally obtained and consequently, the stoichiometric number of the rate-determining step can be determined.[2]

1-4 Rate equations, reaction intermediates and the kinetic structure of chemical reactions

In general, chemical reactions take place through many elementary steps. The chemical potentials of the individual reaction intermediates during the reaction are given by equation (1-3-23) or Fig. 5.

In general, if the overall reaction takes place as illustrated in Fig. 5 with rth step being rate-determining, the difference in the water levels at the narrowest tube (r) is equal to that between L and R. This difference is the water level (free energy) drop causing the overall flow (reaction). All the water levels preceding tube r are equal and the same goes for the levels following tube r.

The diameter of each tube in Fig. 5 can be examined by means of a dye (isotopic tracer). If we put a dye into tank I_3, the water in the tank will rapidly become coloured. The water in tank I_2 will colour at a measurable rate which will depend on the size of tube 3. However, since tube r is very narrow, there will be some delay before the dye reaches tanks I_r, R, ... If the dye is added to tank L, the sequence will be similar, but if it is placed in R, tank I_{n-1} will colour rapidly, where as L, I_1 and I_{r-1} will become coloured at a much slower rate, depending upon the size of tube r. Thus, by measuring the rate at which the dye reaches each tank, the diameter of each tube (or the rate of each step) may be estimated. In this way, isotope tracers may be used to study the kinetic structure of the overall reaction. That is, the elementary steps which make up the overall reaction (or reaction path), their rates and the free energy drop at each step can be elucidated.

If part of the adsorbed species does not participate in the heterogeneous catalysis, or if the reactivity of the adsorbed species is not uniform, the rate of isotope mixing can be used to give information on the heterogeneity of the chemisorbed species. (In this case the concentration of dye throughout the intermediate tank is not uniform.) The isotopic tracer technique as well as the following methods are also applicable to the reaction system in a non-steady state.

The series of water tanks in Fig. 5 can also be treated dynamically to examine the diameters of some of the connecting tubes. The rate of response of each tank to a rapid change of water level in another tank can be followed to study the size of the tubes between them. Alternatively, the water level of one of the tanks may be moved up and down periodically and the propagation rate of the movement to the other tanks studied.

When all the water levels are equal in height and the water in one of the tanks is removed rapidly, the level of each tank will undergo corresponding changes according to the sizes of the connecting tubes. For example, consider a mixture of nitrogen, hydrogen and ammonia in equilibrium over a catalyst surface. If the gas phase ammonia is rapidly removed by a liquid nitrogen trap, then provided that the nitrogen adsorption is the only slow step in the overall ammonia synthesis, the amount of adsorbed nitrogen should decrease quite rapidly (Fig. 6). Using this non-steady state technique, the rate of nitrogen chemisorption as well as the rate of hydrogenation of the chemisorbed nitrogen to form ammonia can be separately measured as a function of the coverage and the pressure of reacting gases, and the mechanism of the

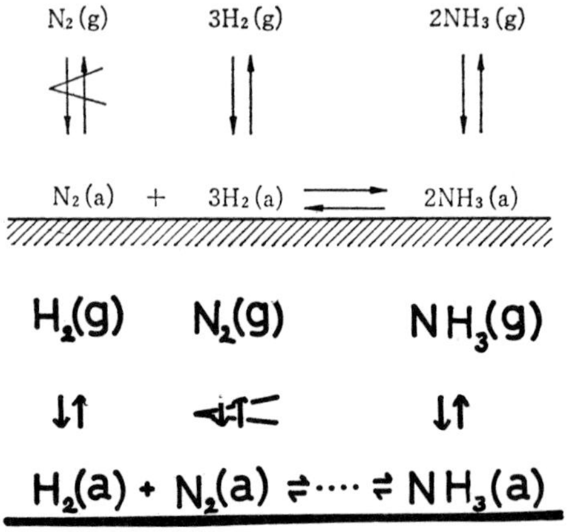

N_2 (g) $3H_2$ (g) $2NH_3$ (g)

N_2 (a) + $3H_2$ (a) ⇌ $2NH_3$ (a)

H_2(g) N_2(g) NH_3(g)

H_2(a) + N_2(a) ⇌ ···· ⇌ NH_3(a)

Fig. 6.

ammonia synthesis may be elucidated. Accordingly, this approach may be used to separate the overall reaction into the simpler processes from which it is built up and to identify the rate-determining step.

If, as illustrated in Fig. 7, a gas molecule, A_2(g) reacts directly with an adsorbed species, B_2(a), to form the reaction product, C(g), without proceeding through A_2(a), the mechanism may be confirmed by removing A_2(g) rapidly from the ambient gas. The overall reaction would stop very rapidly even though A_2(a) and B_2(a) still remained on the surface. Such a non-stationary treatment may also be applied to adsorption measurements under reaction conditions.

It is to be expected therefore, that the kinetic structure of the overall reaction will be able to be elucidated by the application of the dynamic techniques that have been outlined. It should be noted that the existence of a certain adsorbed species on the catalyst surface does not necessarily imply that it is actually a reaction intermediate involved in the reaction sequence. The identification of a reaction intermediate in the adsorbed state can only be carried out by studying its kinetic behaviour in the working state.

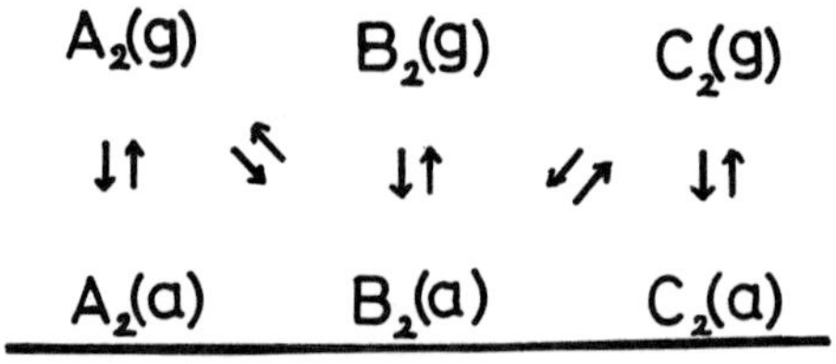

A_2(g) B_2(g) C_2(g)

A_2(a) B_2(a) C_2(a)

Fig. 7.

In homogeneous gas reactions the mechanism of the overall reaction and the real reaction path may be elucidated by studying all of the elementary steps which could possibly be involved in the overall reaction. These elementary steps may be independently examined by other techniques such as, shock waves, flash photolysis and the flow method.

In the case of heterogeneous catalysis, the properties of the surface where the reaction takes place (such as the work function) are influenced to various extents by the adsorption of the reactants, intermediates and products depending upon the properties and amounts of the adsorbed species. The adsorption (or chemical potential) of these adsorbed species is also affected by the location of the rate-determining step, which is also influenced by the properties of the environment where the reaction takes place or by the surface properties of the catalyst. Under these circumstances it is not always sensible to isolate each of the elementary steps of the catalytic reaction as we do for gas reactions, since in general, it is not the properties of the bare catalyst surface that are important, but the properties of the surface on which various surface species are adsorbed. Consequently, the structures, surface concentrations and reactivities of the adsorbed species should be examined during the course of the reaction, along with the rate of the overall reaction.(3)

1-5 Catalysis and chain reactions

Hydrogen and chlorine will undergo thermal reaction to form hydrogen chloride and the reaction can also be initiated photochemically at room temperature. It is generally accepted that the first step is the dissociation of a chlorine molecule to chlorine atoms. The chlorine atoms thus formed react with hydrogen molecules to give hydrogen chloride and hydrogen atoms. The hydrogen atoms react with additional chlorine molecules to give hydrogen chloride and chlorine atoms. The overall reaction (1-5-1) proceeds by repetition of reactions (1-5-3) and (1-5-4). The hydrogen and chlorine atoms which are repeatedly generated and consumed in the course of the overall reaction, are called "chain carriers" and a reaction which proceeds via such chain carriers is called a "chain reaction". Most combustion, explosion and polymerization reactions proceed via a chain reaction mechanism.

$$H_2 + Cl_2 = 2HCl \qquad (1\text{-}5\text{-}1)$$

$$(1) \quad Cl_2 \xrightarrow[h\nu]{\Delta} 2Cl \qquad (1\text{-}5\text{-}2)$$

$$(2) \quad Cl + H_2 \rightarrow HCl + H \qquad (1\text{-}5\text{-}3)$$

$$(3) \quad H + Cl_2 \rightarrow HCl + Cl \qquad (1\text{-}5\text{-}4)$$

Catalysis may also be thought of as a chain reaction. Consider the case where A and B react extremely slowly to form C and D, even though the overall reaction

$$A + B = C + D \tag{1-5-5}$$

is thermodynamically possible. If we add X or Y to the system, and if the following steps take place at considerable rates, X and Y are alternately consumed and regenerated. Every time these cycling processes are repeated,

$$\begin{aligned} &\text{(i)} \quad A + X = C + Y \\ &\text{(ii)} \quad B + Y = D + X \end{aligned} \tag{1-5-6}$$

reaction products are formed and the overall reaction (1-5-5) proceeds at a substantial rate. Here X and Y play an important role in the cycling processes (1-5-6), but do not appear in the stoichiometric equation of the overall reaction. Such species which promote the overall reaction are called "catalysts" for the reaction (1-5-5). They promote reaction (1-5-5) and their role is quite similar to that of the chain carriers, H and Cl in processes (1-5-3) and (1-5-4) of reaction (1-5-1). (A, X, C, Y, B and D in (1-5-6) correspond to H_2, Cl, HCl, H, Cl_2 and HCl, respectively, in (1-5-3) and (1-5-4).) Consequently, catalytic reactions are always chain reactions and the catalyst a chain carrier. None of the elementary steps by themselves shows any features characteristic to catalysis or chain reactions, but the way in which they are combined to form a characteristic cycling process gives the overall reaction a faster rate.†

Accordingly in a catalytic reaction (or chain reaction), if one of the elementary steps in the cycle is very slow, the overall reaction will proceed very slowly, although the chemical potentials of the reactants of the step would be raised correspondingly. If a foreign substance is present which combines strongly with one of the chain carriers (the catalyst) to retard one of the steps in the cycle, it will retard or poison the catalytic reaction.‡

† As catalysis is characterized by the mode of combination of the elementary steps and not by the elementary steps which comprise the overall reaction, the fundamental problems in catalysis are, on the whole, common to those of chemistry in general. New techniques or theories appearing in chemistry are usually applicable to catalysis and new approaches developed in the field of catalysis may be applied to chemistry in general.

‡ The activity of a catalyst depends on the ability to promote the overall reaction, and it varies, of course, with the nature of the overall reaction. Hence, in this sense, it is meaningless to say "the activity of a certain catalyst" without mentioning the reaction to be promoted.

In the field of heterogeneous catalysis, the catalyst phase is different from the reactant phase and in most cases we are concerned with solid catalysts for gaseous reactions, where the solid surface plays an important role. Consider a reaction where a substance A, which is stable in the absence of a catalyst (for example butene or formic acid), isomerizes or decomposes over a catalyst to give compound C.

$$A \rightarrow C \tag{1-5-7}$$

The first step is the interaction between the reactant A and the catalyst and, in most cases, this involves the adsorption of A on the catalyst surface. On adsorption the molecule A is chemically influenced by the catalyst, (for example, it becomes deformed, dissociates, or takes part in an electron transfer process with the catalyst, either as an electron acceptor or electron donor), such that its isomerization or decomposition to form C becomes easier with respect to its free state.

$$\begin{aligned} &\text{(i)} \qquad A + S \rightleftarrows A\text{–}S \\ &\text{(ii)} \qquad A\text{–}S \rightleftarrows C + S \end{aligned} \tag{1-5-8}$$

where S represents the bare catalyst surface or a surface site and A–S represents a molecule of A adsorbed on the surface. The cyclic form (1-5-8) is similar to that of (1-5-6); this can be readily seen by replacing X by S, Y by A–S, and B and D by zero.

If both steps of equation (1-5-8) proceed at a substantial rate, catalytic reaction takes place. In the steady-state, the concentration of the chain carrier stays approximately unchanged;

$$-\mathrm{d(S)}/\mathrm{d}t = 0 = k_{+1}(\mathrm{A})(\mathrm{S}) - k_{-1}(\mathrm{A\text{–}S}) - k_{+2}(\mathrm{A\text{–}S}) + k_{-2}(\mathrm{C})(\mathrm{S}) \tag{1-5-9}$$

where k_{+1}, k_{-1} and k_{+2}, k_{-2} are the forward and backward rate constants for steps (i) and (ii) in reaction (1-5-8), and are assumed to be independent of (A), (S), (A–S) and (C).†

$$(\mathrm{S}) = \frac{k_{-1} + k_{+2}}{k_{+1}(\mathrm{A}) + k_{-2}(\mathrm{C})}(\mathrm{A\text{–}S}) \tag{1-5-10}$$

† In this case the rate constants, k_{+1}, k_{-1}, k_{+2} and k_{-2}, are independent of (A–S) and (S). In many cases, however, the rate constants are in fact dependent on the concentrations of the surface species.

The number of surface sites, (S_0), is the sum of the bare sites (S) and occupied sites (A–S). Accordingly,

$$(S) + (A\text{–}S) = (S_0)$$

$$(A\text{–}S) = \frac{k_{+1}(A) + k_{-2}(C)}{k_{-1} + k_{+2} + k_{+1}(A) + k_{-2}(C)}(S_0)$$

$$(S) = \frac{k_{-1} + k_{+2}}{k_{-1} + k_{+2} + k_{+1}(A) + k_{-2}(C)}(S_0)$$

Thus, the rate of reaction per surface site is given as follows:

$$\frac{1}{S_0}\frac{d(C)}{dt} = \frac{1}{S_0}\left(k_{+2}(A\text{–}S) - k_{-2}(C)\right)$$

$$= \frac{k_{+2}\left(k_{+1}(A) + k_{-2}(C)\right) - k_{-2}(C)(k_{-1} + k_{+2})}{k_{-1} + k_{+2} + k_{+1}(A) + k_{-2}(C)}$$

$$= \frac{k_{+1}k_{+2}(A) - k_{-1}k_{-2}(C)}{k_{-1} + k_{+2} + k_{+1}(A) + k_{-2}(C)} \tag{1-5-11}$$

If k_{-2} is negligible in comparison with other rate constants;

$$\frac{1}{S_0}\frac{d(C)}{dt} = \frac{k_{+1}k_{+2}(A) - k_{-1}k_{-2}(C)}{k_{+1}(A) + k_{-1} + k_{+2}}$$

$$= \frac{k_{+2}\left(\frac{k_{+1}}{k_{-1} + k_{+2}}\right)(A)}{1 + \frac{k_{+1}}{k_{-1} + k_{+2}}(A)} - \frac{\frac{k_{-1}k_{-2}}{k_{-1} + k_{+2}}(C)}{1 + \frac{k_{+1}}{k_{-1} + k_{+2}}(A)} \tag{1-5-12}$$

This rate equation is well known in heterogeneous catalysis as the Langmuir equation, and in the field of enzyme reactions as the Michaelis–Menten equation, generally being expressed as:

$$\text{rate of reaction} = \frac{k(A)}{1 + k'(A)} \tag{1-5-13}$$

1-6 A kinetic study of the reaction, $H_2 + Br_2 = 2HBr$

As an example of the kinetic treatment of chain reactions, the kinetics of the reaction between hydrogen and bromine will be examined in more detail.

$$H_2 + Br_2 = 2HBr \tag{1-6-1}$$

The rate equation for this reaction has been experimentally determined as:

$$\frac{d(HBr)}{dt} = \frac{k(H_2)(Br_2)^{\frac{1}{2}}}{1 + k'\dfrac{(HBr)}{(Br_2)}} \tag{1-6-2}$$

The temperature dependence of the rate constant k gives an activation energy of 40·2 kcal/mol, while the other constant in the equation, k', is practically independent of temperature, being 0·116 and 0·112 at 25° and 300°C, respectively.

To explain the rate equation (1-6-2), the following reaction sequences have been proposed:

(a) $Br_2 = 2Br - 46{\cdot}0\ \text{kcal}\dagger;\ K_{Br} = (Br)/(Br_2)^{\frac{1}{2}} = 9{\cdot}9 \times 10^{-13}\,(\text{atm})^{\frac{1}{2}}$

(1-6-3a)

(b) $Br + H_2 = HBr + H - 16{\cdot}7\ \text{kcal};\ K_p^b = 8{\cdot}8 \times 10^{-15}$ (1-6-3b)

(c) $H + Br_2 \rightarrow HBr + Br + 41{\cdot}4\ \text{kcal};\ K_p^c = 1{\cdot}8 \times 10^{33}$ (1-6-3c)

where K_{Br}, K_p^b and K_p^c are the equilibrium constants of steps (a), (b) and (c), respectively at 298 K.

The concentrations of hydrogen and bromine atoms, H and Br, in the reacting system are extremely low during the course of the reaction and the steady-state approximation is applied as follows:

$$\begin{aligned} d(Br)/dt &= k_{+1}(Br_2) - k_{-1}(Br)^2 - k_{+2}(H_2)(Br) + k_{-2}(HBr)(H) \\ &\quad + k_{+3}(H)(Br_2) \\ &= 0 \end{aligned} \tag{1-6-4}$$

† In the thermal reaction it is better to write this step as: $Br_2 + M = 2Br + M - 46{\cdot}0$ kcal, where M is a third body.

$$d(H)/dt = k_{+2}(Br)(H_2) - k_{-2}(HBr)(H) - k_{+3}(H)(Br_2)$$
$$= 0 \tag{1-6-5}$$

where k_{+1}, k_{-1}, k_{+2}, k_{-2} and k_{+3} are the forward and backward rate constants of steps (a), (b) and (c) in equation (1-6-3), respectively.

Thus, all possible steps that can produce and consume Br or H are balanced. Consequently,

$$(Br) = (k_{+1}/k_{-1})^{\frac{1}{2}}(Br_2)^{\frac{1}{2}} = K_{Br}(Br_2)^{\frac{1}{2}} \tag{1-6-6}$$

$$(H) = \frac{k_{+2}(Br)(H_2)}{k_{-2}(HBr) + k_{+3}(Br_2)} = \frac{k_{+2}(k_{+1}/k_{-1})^{\frac{1}{2}}(Br_2)^{\frac{1}{2}}(H_2)}{k_{-2}(HBr) + k_{+3}(Br_2)} \tag{1-6-7}$$

As the rate of HBr formation is the sum of the rates of all the steps which produce HBr minus these that consume HBr,

$$d(HBr)/dt = k_{+2}(H_2)(Br) - k_{-2}(HBr)(H) + k_{+3}(H)(Br_2)$$
$$= \frac{2k_{+2}(k_{+1}/k_{-1})^{\frac{1}{2}}(H_2)(Br_2)^{\frac{1}{2}}}{1 + (k_{-2}/k_{+3})(HBr)/(Br_2)} \tag{1-6-8}$$

The form of rate equation (1-6-8) is in good agreement with that of (1-6-2) and k and k' correspond to the following ratios of the rate constants:

$$k = 2k_{+2}(k_{+1}/k_{-1})^{\frac{1}{2}} = 2k_{+2}(K_{Br}) \tag{1-6-9}$$

$$k' = k_{-2}/k_{+3} = 0{\cdot}12 \tag{1-6-10}$$

If the activation energies for each of the rate constants are designated in a similar manner as E_{+1}, E_{-1}, E_{+2}, E_{-2}, and E_{+3}, equation (1-6-10) shows that

$$E_{-2} = E_{+3} \tag{1-6-10'}$$

since k_{-2}/k_{+3} is temperature independent.

The temperature dependence of K_{Br} gives the following relation:

$$E_{+1} - E_{-1} = 46{\cdot}1 \text{ kcal/mol}$$

Since the backward reaction of (1-6-3a) is the recombination reaction of atoms, its activation energy should be practically zero,

$$E_{-1} = 0, E_{+1} = 46{\cdot}1 \text{ kcal/mol}$$

Equation (1-6-9) leads to the equation:

$$40{\cdot}2\ \text{kcal} = E_{+2} + \tfrac{1}{2}(E_{+1} - E_{-1})$$

$$E_{+2} = 40{\cdot}2 - 23{\cdot}0 = 17{\cdot}2\ \text{kcal}$$

Consequently, as $E_{+2} - E_{-2} = 16{\cdot}7$ kcal/mol,

$$E_{-2} = 0{\cdot}5\ \text{kcal/mol}$$

From equation (1-6-10′)

$$E_{+3} = 0{\cdot}5\ \text{kcal/mol}$$

In this way all of the activation energies for the five rate constants may be estimated.†

The concentration of hydrogen atoms in the steady-state of the reaction, $(\text{H})_{\text{SS}}$, is given by equation (1-6-7).

$$(\text{H})_{\text{SS}} = \frac{k_{+2}(k_{+1}/k_{-1})^{\frac{1}{2}}(\text{H}_2)(\text{Br}_2)^{\frac{1}{2}}}{k_{-2}(\text{HBr}) + k_{+3}(\text{Br}_2)}$$

$$= \frac{(k_{+2}/k_{+3})(k_{+1}/k_{-1})^{\frac{1}{2}}(\text{H}_2)/(\text{Br}_2)^{\frac{1}{2}}}{1 + (k_{-2}/k_{+3})(\text{HBr})/(\text{Br}_2)} \qquad (1\text{-}6\text{-}11)$$

As k_{-2}/k_{+3} is 0·12 from equation (1-6-10), the second term in the denominator may be neglected at the beginning of the reaction where $(\text{HBr}) \geq (\text{Br}_2)$, especially when we are estimating $(\text{H})_{\text{SS}}$.
Thus,

$$(\text{H})_{\text{SS}} = \frac{k_{+2}}{k_{+3}}\left(\frac{k_{+1}}{k_{-1}}\right)^{\frac{1}{2}} \frac{(\text{H}_2)}{(\text{Br}_2)^{\frac{1}{2}}} \qquad (1\text{-}6\text{-}12)$$

If the equilibrium constant for hydrogen dissociation is expressed by K_{H}, $K_{\text{H}} = (\text{H})_e^2/(\text{H}_2)$:

$$\frac{(\text{H})_{\text{SS}}}{(\text{H})_e} = \frac{k_{+2}}{k_{+3}}\left(\frac{(\text{H}_2)}{(\text{Br}_2)}\right)^{\frac{1}{2}} \frac{K_{\text{Br}}}{K_{\text{H}}^{\frac{1}{2}}}$$

$$= \frac{k_{-2}}{k_{+3}} K_p^b \left(\frac{(\text{H}_2)}{(\text{Br}_2)}\right)^{\frac{1}{2}} \frac{K_{\text{Br}}}{K_{\text{H}}^{\frac{1}{2}}} \qquad (1\text{-}6\text{-}13)$$

† These activation energies have been independently confirmed by other experiments such as shock tube, which further supports the reaction mechanism.

As we know the values of k_{-2}/k_{+3} (0·12), K_p^b, K_{Br}, and $K_H(2{\cdot}0 \times 10^{-36}(\text{atm}))$, $(H)_{SS}/(H)_e$ may be calculated at 298 K for $(H_2) = (Br_2)$ to give:

$$(H)_{SS}/(H)_e = 5{\cdot}2 \times 10^8 \tag{1-6-14}$$

The hydrogen atom concentration in the steady-state of the reaction is very much higher than that given by the dissociative equilibrium of H_2. This is because the hydrogen atoms in the reaction system are not supplied by the dissociation of hydrogen molecules, but through the following process:

$$Br + H_2 = HBr + H \tag{1-6-3b}$$

The bromine atom in the reaction system, on the other hand, is in dissociation equilibrium with molecular bromine as given by equation (1-6-6), and the ratio $(Br)_{SS}/(Br)_e$ is unity during the reaction.

The equilibrium constant of reaction (1-6-15)

$$\tfrac{1}{2}Br_2 + H_2 = HBr + H \tag{1-6-15}$$

at 298 K can be determined from the values of K_{Br} and K_p^b in equations (1-6-3a) and (1-6-3b) as $0{\cdot}88 \times 10^{-26}$ (atm^{-1}). Consequently, if $(HBr) = (H_2) = (Br_2) = 1/3$ atm, the concentration of hydrogen atoms that would be in equilibrium with the gas mixture, $(H)^{\circ}_{eq}$, can be calculated as $0{\cdot}51 \times 10^{-26}$ atm($(H)_e$ in dissociation equilibrium with 1/3 atm (H_2) is $1{\cdot}2 \times 10^{-36}$ atm, a much lower pressure than $(H)^{\circ}_{eq}$, while $(H)_{SS}$ is $6{\cdot}1 \times 10^{-28}$ atm). Accordingly, the free energy drop for step (1-6-3b) at 298 K is

$$-RT\ln(6{\cdot}1 \times 10^{-28}/0{\cdot}5 \times 10^{-26}) = 1{\cdot}25 \text{ kcal}$$

The total free energy drop for the overall reaction (1-6-1) at $(H_2) = (Br_2) = (HBr) = 1/3$ atm is 25·44 kcal and that for step (1-6-3a) is zero. Therefore, the free energy drop over step (1-6-3b) corresponds to approximately 5% of that of the overall reaction and the main portion of the drop occurs in the last step of the reaction, that is, in step (1-6-3c).

This is understandable, as the free energy drop is associated with the ratio of the forward and backward reaction rates, as shown in equation (1-3-18). In step (1-6-3b) both rates are roughly comparable, whereas in the third step, (1-6-3c), the backward rate is negligible in comparison to the forward rate. This is because the step is a very exothermic reaction and the backward process will have an activation energy greater than 41 kcal/mol.

In this way, by knowing the rate constants of all the elementary steps, we can determine the reaction path of the overall reaction and can estimate the concentrations of the reaction intermediates and the free energy drops for each of the elementary steps during the course of the reaction. The kinetic structure of the overall reaction may thus be established. It has also been demonstrated that the main decrease in free energy for the overall reaction, (1-6-1), takes place at the third step, (1-6-3c), and this may be considered as the main rate-determining step of the overall reaction.

It is to be noted here that the forward rate constant of the rate-determining step, k_{+3}, is not the smallest one in the forward rate constants of the three steps, (a), (b) and (c), but is, in fact, for this example, the largest. The activation energy of k_{+1} is estimated to be 46·1 kcal/mol, that of k_{+2} 17·2 kcal/mol, and that of k_{+3} 0·5 kcal/mol. The differences in the frequency factors, or pre-exponential factors of the rate constants of these simple gas reactions are much smaller than the differences in the exponential factors, $\exp(-E/RT)$.† Consequently,

$$k_{+1} \ll k_{+2} \ll k_{+3} \tag{1-6-16}$$

The third step, which, in this case, has the largest forward rate constant is the main rate-determining step in this particular overall reaction. The rate-determining step which determines the overall kinetics is, accordingly, not the step which has the smallest rate constant.

If the third step is the only rate-determining step in the overall reaction and if all the other steps are in equilibrium during the course of the reaction,

$$\begin{aligned} \mathrm{d(HBr)/d}t &= k_{+3}(\mathrm{H})(\mathrm{Br_2}) \\ &= k_{+3}\frac{k_{+2}}{k_{-2}}\frac{k_{+1}}{k_{-1}}\frac{(\mathrm{H_2})(\mathrm{Br_2})^{\frac{3}{2}}}{(\mathrm{HBr})} \end{aligned} \tag{1-6-17}$$

as $(\mathrm{H}) = K_p^b K_{\mathrm{Br}}^{\frac{1}{2}}(\mathrm{Br_2})^{\frac{1}{2}}(\mathrm{H_2})/(\mathrm{HBr})$. This implies that the one in the denominator of equation (1-6-8) has become small compared to the second term.

We can use a comparison between reaction (1-6-1) and the analogous hydrogen iodine reaction as an example of another characteristic feature of chemical reactions in general. In reaction (1-6-1) the concentration of bromine atoms is determined by the dissociative equilibrium constant of bromine. For the reaction of hydrogen with iodine to form hydrogen iodide in the gas phase,

$$H_2 + I_2 = 2HI \tag{1-6-18}$$

† The pre-exponential factors of k_{+2}, and k_{+3} are as follows: $A_{+2} = 9{\cdot}31$, $A_{+3} = 9{\cdot}83$ litres/mol sec.

the thermodynamic data (enthalpy changes) at 25°C are as given in Table IV. The Table shows that when X = I, the second step is very endothermic, and that the dissociation of the iodine molecule is much easier than that of bromine.

Table IV. The standard enthalpy changes (kcal/mol), ΔH_0^0

	Cl	Br	I
$\frac{1}{2}X_2 = X$	29·0	23·0	17·7
$X + H_2 = HX + H$	1·0	16·7	32·9
$H + X_2 = HX + X$	−45·1	−41·4	−36·0

As the second step is so endothermic when X = I, the activation energy for k_{+2} will be large for the HI formation reaction. This will mean that even though the concentration of iodine atoms is much higher than the concentration of bromine atoms at the same temperature, the chain mechanism will not be valid for HI formation at low temperatures. This emphasizes the point that the presence of a species which is not one of the reactants or products does not necessarily mean that it is a reaction intermediate.

As the activation energy of the chain mechanism is higher than that of the non-chain reaction, reaction (1-6-18) should proceed through the chain mechanism at much higher temperatures. Actually, 10%, 27% and 95% of the overall reaction (1-6-18) proceeds via the chain mechanism at 633, 738 and 800 K, respectively.[4]

The reactions between hydrogen and the halogens are typical examples of chain reactions for which the reaction mechanisms have been studied in detail. In the case of heterogeneous catalysis the fundamental approach to elucidating the reaction mechanism is, in many respects, similar. However, although the rates of surface reactions are studied separately to determine rate constants, these investigations are not always directly associated with the eludication of reaction mechanisms. Even if we detect the most abundantly adsorbed species on the catalyst surface by some spectroscopic technique, it does not necessarily lead to any conclusions on the reaction intermediate, unless the chemical potentials of the adsorbed species are examined in a quantitative manner. In particular, as has been mentioned previously, the nature of the catalyst surface is sensitive to the presence of adsorbed species, and elementary reactions on the surface should consequently be studied under the reaction conditions.

References

1. J. Horiuti, *Proc. Japan Academy*, **29**, 160, 164 (1953); *Adv. in Catalysis*, **IX**, 339 (1957); *Chem. and Ind.* (*Chem. Soc. Japan*),), **9**, 355 (1956).
2. K. Tanaka, "Proc. Third Internl. Cong. on Catalysis, I" p. 676 (1964).
3. K. Tamaru, *Adv. in Catalysis*, **15**, 65 (1964).
4. J. H. Sullivan, *J. Chem. Phys.* **30**, 1292 (1959).

Chapter 2

Adsorption

2-1 Physical and chemical adsorption

When charcoal is exposed to a gas or a vapour, adsorption takes place and the concentration of gas or vapour molecules at the surface of the charcoal becomes higher than that in the ambient gas. The nature of the attractive force between the solid surface and the adsorbed molecules depends upon the properties of the surface and the adsorbed species. If the force is physical in nature, it is called physical adsorption or physisorption, and an example of physical attractive force is the van der Waals force between molecules in a liquid. In such cases, the strength of the cohesive force depends upon the physical properties of the gases and vapours to be adsorbed, and is generally higher for gases (vapours) with higher boiling points.

The properties of solid surfaces can be markedly different from those of the bulk solid because the surface atoms are unsaturated. For example, diamond surfaces may conduct electricity although the bulk solid is renowned as an insulator. The unsaturated bonds present at the surface provide sites for the adsorption of gas molecules. This type of adsorption, involving a chemical reaction between the adsorbed molecules and the surface atoms, is called chemisorption.

The heat evolved by physisorption is often of the same order of magnitude as the heat of liquefaction of the adsorbed species. For chemisorption, the heat evolved is of the same order of magnitude as that of chemical reactions and in many cases, the heats of chemisorption are one or two orders of magnitude higher than those of physical adsorption. In general, the chemical bonds formed between the surface and the adsorbed molecules in chemisorption are stronger than the physical forces involved in physisorption.

Physical adsorption can be considered as a condensation phenomenon and because it is nonspecific, multilayer adsorption can take place. An example of the application of the nonspecific nature of physical adsorption is the determination of surface areas by the physical adsorption of nitrogen, for example, at low temperatures. (See the BET adsorption isotherm in Section 2–10.) The adsorption isotherm can be analysed, by the BET method in order to determine the amount of gas required for monolayer formation (V_m). Since a molecule of physisorbed gas is not associated with any particular surface atom, it occupies an area (molecular size) independent of the nature of the solid surface, and the surface area of the solid can therefore be determined.

Chemisorption continues until all the unsaturated surface valencies are satisfied and this of course corresponds to the adsorption of a monolayer. Since chemisorption is similar to chemical reaction between the adsorbed gas and the solid surface, it is perhaps not surprising that chemisorption is highly specific. Nitrogen is chemisorbed at moderate temperatures on a limited number of solid surfaces, iron and tungsten for example, but not by such metals as zinc, copper, silver or platinum. Hydrogen is chemisorbed by platinum, palladium, iron and tungsten at room temperature, but not by silver gold, cadmium, lead or zinc.

Various evaporated metals and semi-metals may be classified into several groups depending upon the gases adsorbed at room temperature, in the manner shown in Table I.[1] The strength and the rate of adsorption also

Table I. Classification of metals and semi-metals based on adsorption properties (A indicates adsorption, NA no adsorption)

Group	Metals	Gases						
		O_2	C_2H_2	C_2H_4	CO	H_2	CO_2	N_2
A	Ca, Sr, Ba, Ti, Zr, Hf, V, Nb, Ta, Cr, Mo, W, Fe[a], (Re)	A	A	A	A	A	A	A
B_1	Ni, (Co)	A	A	A	A	A	A	NA
B_2	Rh, Pd, Pt, (Ir)	A	A	A	A	A	NA	NA
C	Al, Mn, Cu, Au	A	A	A	A	NA	NA	NA
D	K	A	A	NA	NA	NA	NA	NA
E	Mg, Ag[a], Zn, Cd, In, Si, Ge, Sn, Pb, As, Sb, Bi	A	NA	NA	NA	NA	NA	NA
F	Se, Te	NA	NA	NA	NA	NA	NA	NA

[a] The adsorption of N_2 on Fe is activated, as is the adsorption of O_2 on Ag films sintered at 0°C.
( ) Metal probably belongs to this group, but the behaviour of films is not known.

depend upon the crystal faces exposed on the surface as revealed by FEM and FIM techniques (see Section 2-3).

The fact that solids can catalyse gas reactions has been recognized for quite some time. It is more than 150 years since Davy discovered the catalytic activity of platinum in the combustion of coal gas, hydrogen and alcohol, and as early as 1834, Faraday associated this type of catalytic activity with the adsorption of gases on the solid catalysts. In both chemical and physical adsorption, the density of gas or vapour molecules in the adsorbed layer is higher than that in the ambient gas, corresponding to a higher pressure. However, the acceleration of chemical reactions by catalysts is generally not a result of this higher concentration, but is more often due to chemical factors, such as the dissociation or distortion of the gas molecules when they become chemisorbed on a solid surface. These chemical effects involved in chemisorption are the reason that some solid surfaces can participate in catalytic reactions as depicted in equation (1-5-8).

An example of the role of adsorbed species in heterogeneous catalysis is the effect of metal catalysts on the reaction between hydrogen and oxygen. When, for example, metallic copper is introduced into a mixture of the gases, oxygen is chemisorbed to form a surface oxide and this is subsequently reduced by the hydrogen. The reaction in the presence of copper will occur at a considerable rate in temperature ranges where the gas reaction is still negligible because the oxygen in the chemisorbed oxide layer has increased reactivity towards hydrogen. Similarly, on a palladium surface, chemisorbed hydrogen and oxygen will react even at room temperature. In this sense, the reactivity of surface species under the reaction conditions is of primary importance in heterogeneous catalysis.

2-2 The preparation of clean surfaces

Since the presence of impurities or preadsorbed gases on a surface will markedly alter subsequent adsorption behaviour, the first step in a study of the adsorption of gases or vapours on a solid is the preparation of a clean surface.

There are a number of possible sources of surface contamination. The specimen may possess surface impurities resulting from its pretreatment, for example the diffusion of impurities such as sulphur from the bulk of the specimen to the surface. Other sources of contamination can come from the ambient atmosphere such as grease, oil, mercury and water vapours and carbon monoxide.

Some contaminants may be evaporated or desorbed from the surface by heating the specimen (e.g. tungsten) to just below its melting point under ultrahigh vacuum, some such as carbonaceous residues may be removed by treating the sample with oxygen. If oxygen itself is the contaminant it may be removed by hydrogen treatment. Surfaces may also be cleaned by bombardment with fast-moving positive ions of a nonadsorbing gas such as argon.

Another method of preparing clean surfaces is to produce a new surface in the absence of contaminants. The evaporated film technique is frequently employed, where a metal film is produced by evaporating metal from an electrically heated filament under ultrahigh vacuum. Binary alloy films may also be prepared by simultaneous evaporation from two heated filaments. The formation of new surfaces by cleavage in vacuo is also sometimes used.

The frequency of collision of ambient gas molecules with a solid surface is represented by the following equation from the kinetic theory of gases.

$$Z = \frac{n}{4}\left(\frac{8kT}{\pi m}\right)^{\frac{1}{2}} = 2{\cdot}67 \times 10^{25} P/(MT)^{\frac{1}{2}}$$

where Z is the number of gas molecules colliding with unit area of solid surface per second, n, the number of gas molecules per unit volume, m, the mass of a molecule, M, its molecular weight, P, the pressure of the ambient gas in atmospheres, k, the Boltzmann constant and T, the temperature of the gas. Obviously, in order to maintain a clean surface, the partial pressures of contaminants which may be strongly chemisorbed, such as carbon monoxide and water vapour, should be as low as possible. Assuming that every collision leads to chemisorption and that a saturated adsorption layer is formed by the adsorption of 10^{15} molecules per square centimetre of surface area (which is the order of magnitude of the density of surface atoms) it would take three seconds to form a monolayer at a pressure of 10^{-6} torr and one hour at 10^{-9} torr for $M = 30$ and $T = 300$.†

Adsorption measurements can be carried out by measuring either volumetrically or gravimetrically, the amounts of gases and vapours adsorbed under various pressures. If the surface area of the specimen is $10\,m^2/g$, saturated monolayer formation would correspond to 10^{20} molecules/g or 4 ml (stp)/g, assuming 10^{15} molecules/cm^2. If the molecular weight of the adsorbing gas is 30, the weight increase of the specimen would be 5 mg/g. A microbalance can readily be used to determine the amount adsorbed, and even adsorption on a single crystal face has been measured by this method.

Many other techniques that measure changes in the physical properties of the solid due to adsorption may also be employed, examples being the changes in the work function, accommodation coefficient, electrical conductivity and magnetic susceptibility of the sample.

† It is accordingly important in the experiments on solid surfaces to have ultrahigh vacuum equipment in order to keep the specimen surfaces clean from contamination.

2-3 Field-emission and field-ion microscopy[2, 3]

In a field-emission microscope, a strong field, of the order of 5×10^7 V cm^{-1}, is applied to a sharp tip of a single crystal of metal (Fig. 1). The radius of curvature at the end of the tip is around 1000–3000 Å (see Fig. 2). Under the high electric field, the potential barrier at the surface is lowered such that the

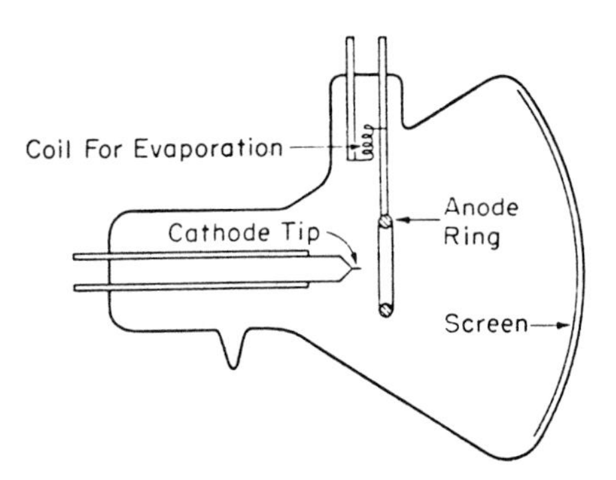

Fig. 1. Diagram of field-emission microscope.

electrons approaching the surfaces of the tip from inside the metal may tunnel the barrier and be emitted. The emitting surface is surrounded by a hemispherical fluorescent screen which may behave as the anode. The emitted electrons hit the screen and the resultant pattern on the screen is a magnified image of the emission from the tip. The extent of magnification (10^5–10^6) is given by the ratio of the distance between the end of the tip and the screen and the radius of curvature of the tip. Areas of the tip with low work functions show up as bright areas on the screen, while those with high work functions appear

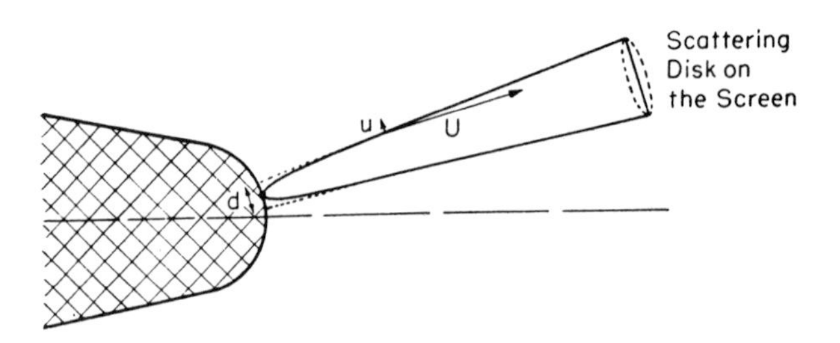

Fig. 2. Scattering disk from a single emitting point on the tip.

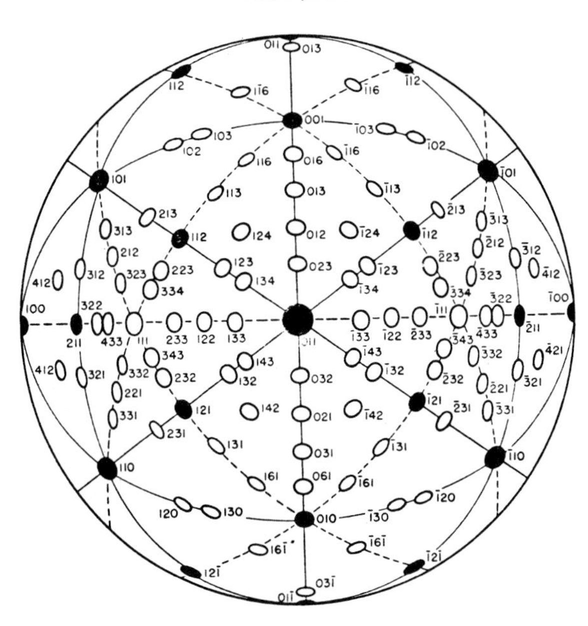

Fig. 3. Map of crystallographic directions.

as dark regions. The picture on the screen essentially represents a work function map of the tip surface of tungsten (see Figs 3 and 4).[3]

When chemisorption takes place on the tip, it changes the work function of the surface. Hence the behaviour of the adsorbed species on the various

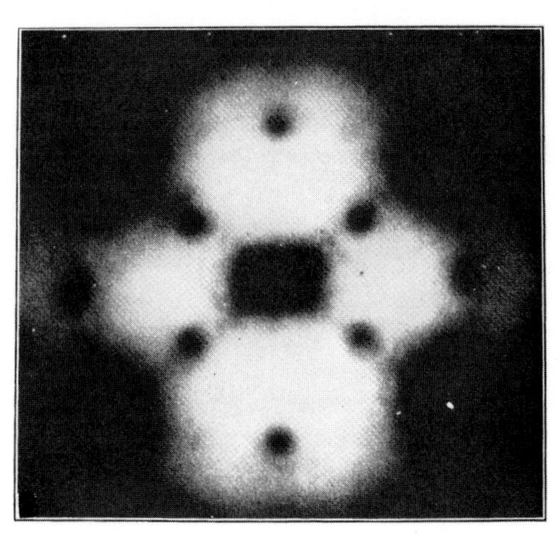

Fig. 4. Field-emission pattern of a clean tungsten tip corresponding to Fig. 3.[3]

exposed crystal faces at the tip is reflected on the screen as changes in the work function. If chemisorption results in a higher work function, the intensity of the pattern decreases. In other words, electronegative adsorbed species will darken the pattern on the screen and vice versa. Accordingly, relative rates of adsorption on different crystal faces may be measured from intensity changes on the screen. The intensity of the various regions of the screen pattern may be determined more accurately by counting the number of electrons by means of a microelectrometer. In some cases it is also possible to examine the mobility of the adsorbed species by this FEM technique.

The resolution of the FEM is limited to about 20 Å as the kinetic energy of an electron emitted from the surface has three degrees of freedom. This gives

Fig. 5. Platinum image (imaged in He, 21 K).[4]

rise to electrons with velocities tangential to the tip surface. In a field-ion microscope, an inert gas such as helium is introduced into an apparatus similar to the FEM. However, with field-ion microscopy (FIM), the tip is made the anode and the pattern on the screen is produced by positive ions of the inert gas. Under the influence of a strong electrostatic field, helium atoms donate electrons to the tip by the overlap of a filled helium orbital with a vacant orbital lobe of a surface metal atom. The image on the screen accordingly gives the distribution of vacant orbital lobes on the tip surface showing the array of the surface atoms of the tip. This technique has a resolution of the order of atomic dimensions.

By means of FEM and FIM techniques, various information on the surface species may be obtained such as, the presence of dislocations or vacancies among the surface atoms, the distribution of vacant orbital lobes of the surface atoms (from the brightness of the various crystal faces in the FIM patterns) (Fig. 5),[(4)] and the presence of foreign atoms on the metal surface. The mobility, adsorption and desorption of the adsorbed species can also be studied and information on the activation energies for desorption and migration and the binding energy of chemisorbed species can be obtained.

The atom-probe FIM method is a technique for analysing ions emitted from various crystal faces on the tip by means of a time-of-flight mass-spectrometer.[(5)] When oxygen is adsorbed on an Fe tip, the emitted ions are FeO^+, FeO_2^+, FeO^{2+} and $FeOH^{2+}$, demonstrating the formation of a surface oxide layer. When Ne or He was used as the image gas, the ions emitted from a Rh tip at 21 K were $RhNe^+$, $RhNe^{2+}$, $RhNe^{3+}$, $RhNe_2^{2+}$, $RhNe^+$ and $RhHe^{2+}$ showing the formation of surface complexes and the weakening of the Rh—Rh bonds in the surface layer. The chemisorption of more stable species such as nitrogen and hydrogen can also be demonstrated by observing the ions RhN^{2+}, RhN^{3+}, and RhH_2^{2+} in a similar manner.†

2-4 Low-energy electron diffraction (LEED)[(7)]

A typical apparatus for LEED studies is shown schematically in Fig. 6. A low-energy electron beam (1–500 eV) is produced by heating a metal or oxide cathode under vacuum, directed by an accelerating potential, and then focused electrostatically on the single crystal surface under study. The surface being examined should be clean and the LEED chamber has to be evacuated to an ultrahigh vacuum of less than 10^{-8} torr.

† The surface reactions which take place on each of the crystal faces, or surface species, may be examined by adding high-voltage pulse and desorbing the surface species, which may be analysed by a time-of-flight mass-spectrometer.[(6)]

Most of the electrons are either inelastically scattered, losing energy in the scattering process, or absorbed by the crystal. However, some of the electrons are elastically back-scattered or back-diffracted by the solid surface. The inelastically scattered ones may be separated from the elastically scattered electrons by means of a retarding potential applied to a grid that repels all the

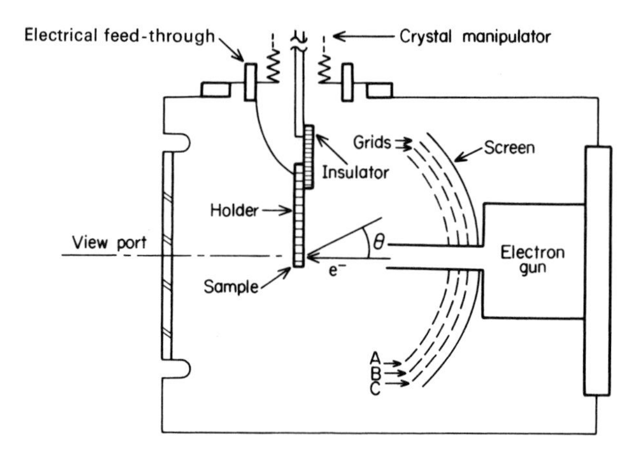

Fig. 6. Scheme of the low-energy electron diffraction apparatus.

electrons that have lost energy in the scattering process. Accordingly, only the electrons back-scattered pass through the retarding field. These are then accelerated by a positive potential onto the spherical fluorescent screen where they display diffraction patterns. Alternatively, a microelectrometer may be employed to count the number of the electrons diffracted to different points. In this fashion, the arrangement of surface atoms may be determined.

The diffraction patterns obtained, however, cannot always be interpreted as a single pattern and may be due to several different ordered surface atom arrangements. Consequently, the intensities of the diffraction spots also have to be examined to give a better insight into the surface arrangements.

Diffraction patterns from Pt(100) surface indicate that an adsorbed ethylene layer has the arrangement as shown in Fig. 7.[8] It has been demonstrated that when unsaturated hydrocarbons, with the exception of isobutylene, are chemisorbed on a Pt(111) surface, a $C(2 \times 2)$ surface structure is formed with unit cell dimensions twice as large as those of the substrate unit cell, suggesting that the adsorbed species occupy every other lattice site on the substrate.

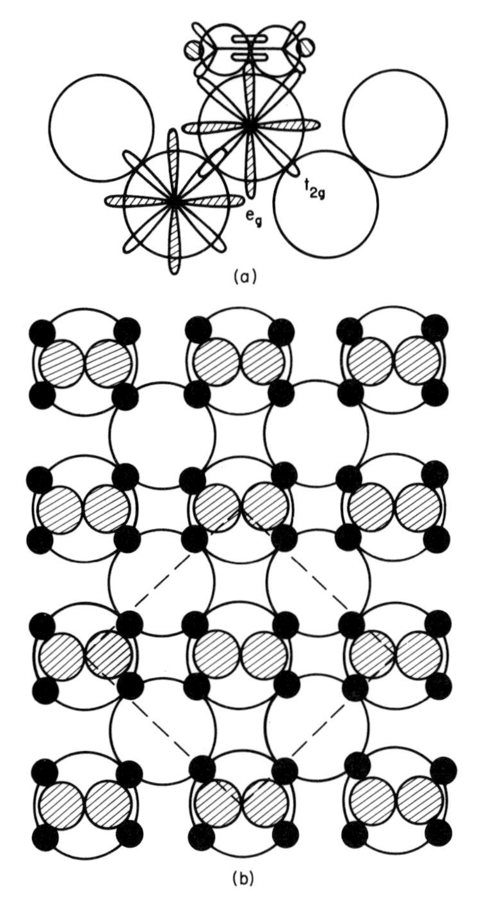

Fig. 7. (a) Bonding of ethylene to the Pt(100) surface. (b) The Pt(100)—C(2 × 2)—(C_2H_4) surface structure.[8]

2-5 Electron spectroscopic techniques

2-5-1 *Electron energy loss spectroscopic method* (*ELS*)
In the ELS method the energy loss spectra of the inelastically scattered electrons are studied in order to obtain information on the energy states involved at the surface. When electrons impinge on a solid surface, the energy distribution of the scattered electrons is as shown in Fig. 8. The sharp peak I is due to the elastically scattered electrons which give the LEED diffraction pattern. The electrons which appear in the range marked II give information

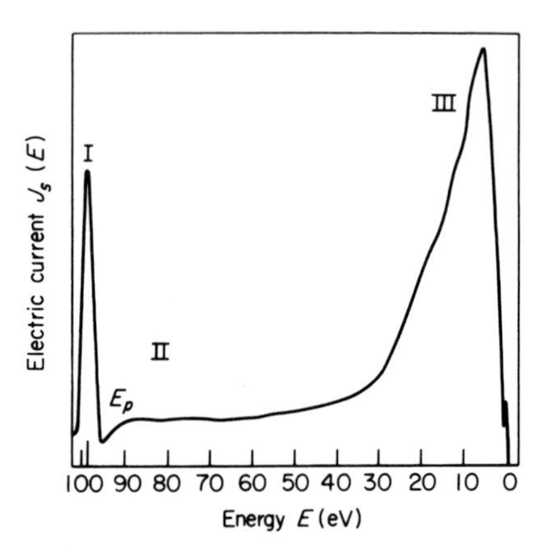

Fig. 8. Energy distribution of secondary electrons from a clean W(110) surface.

on the energy levels at the surface, while those in range III are the true secondary electrons. Vibrational and electronic excitations generally take place at 0·1–1 eV and 1–10 eV, respectively, and consequently it is necessary to have a high resolution spectrum using a monochromatic electron beam. Surface and bulk plasma losses may also be studied.

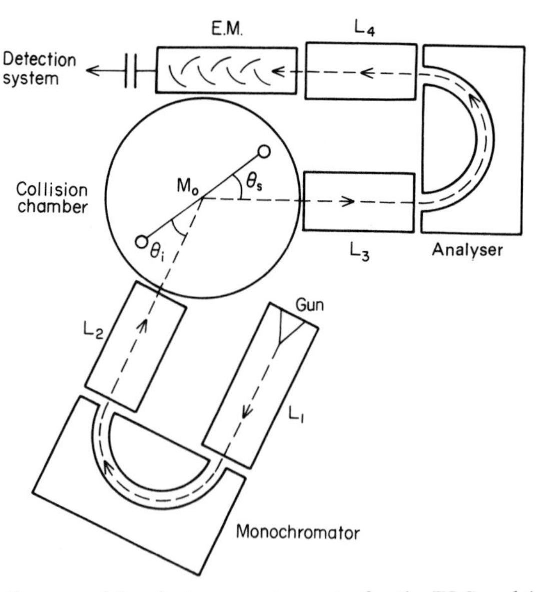

Fig. 9. The block diagram of the electron spectrometer for the ELS and AES for a solid sample.

The apparatus for the ELS method is shown in Fig. 9. Hemispherical electrostatic fields monochromatize and analyse the electron energies and the resolution may reach as high as 40 meV with this technique. Propst and Piper[9] used a monochromatic electron beam (energy of the incident electrons 4·5 eV; resolution 50 meV) and studied the chemisorption of H_2, N_2, CO and so forth on the (100) face of tungsten. The vibrational spectra of the surface species formed when hydrogen and nitrogen were adsorbed are shown in Fig. 10. When hydrogen was preadsorbed and nitrogen subsequently introduced,

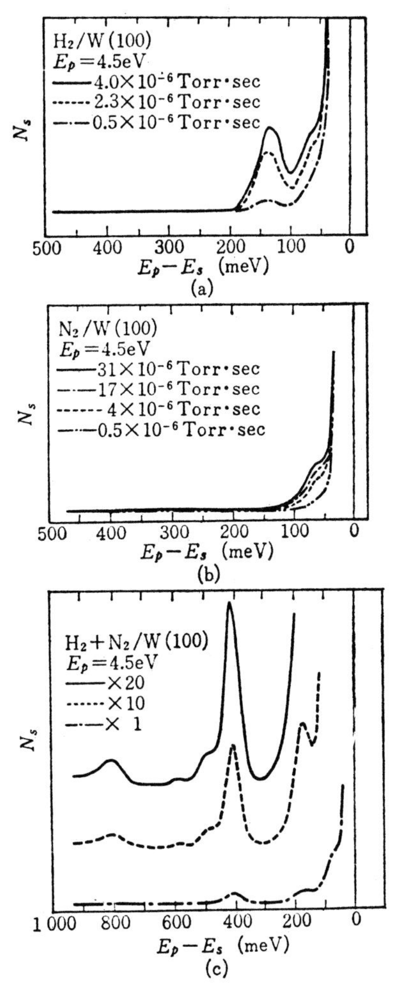

Fig. 10. Energy-loss spectra (W(100) surface).[9]

the spectrum changed as shown in the figure, indicating that the adsorbed hydrogen was replaced by nitrogen. When hydrogen was introduced over a surface on which nitrogen had been preadsorbed, no changes were detected. When a mixture of the two gases was introduced and an ion gauge was operated, two peaks at 404 and 174 meV were observed, which are assigned to NH stretching and NH deformation vibrations. Such vibrational spectra demonstrate the possibility of studying the structure and dynamic behaviour of surface species.

2-5-2 *Auger electron spectroscopy* (*AES*)

If an atom is ionized in an inner shell by collision with a moderate energy beam of electrons (1000–5000 eV) or an X-ray photon, it can relax to a state of lower energy by a process in which an electron from an outer orbital fills the vacancy and the excess energy released by the process is used for secondary electron emission. The energy of this secondary electron emission is characteristic of the atom involved, and Fig. 11 shows an example of an AES spectrum.[10]

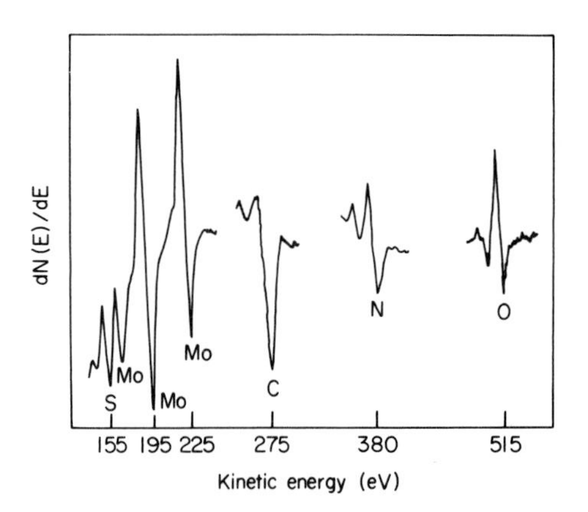

Fig. 11. Differentiated Auger peaks from a molybdenum surface. S—Sulphur, Mo—molybdenum, C—carbon, N—nitrogen and O—oxygen.[10]

Analysis of the energy of these secondary (Auger) electrons can provide us with qualitative and quantitative information about the chemical composition of surfaces. The peaks in the energy spectra, which appear at characteristic energies, identify various surface atoms and the surface concentrations can be related to the intensities of the peaks. The energies of the Auger electrons are independent of the incident electron energy. As there can be considerable interaction between the Auger electrons and the solid, only those electrons originating at the surface retain the characteristic energies.

LEED apparatus may be used for Auger electron spectroscopy, but higher accuracy can be obtained with apparatus similar to that in Fig. 9, where a hemispherical energy analyser is employed. AES has high sensitivity and surface species at less than 1% of a monolayer can be qualitatively and

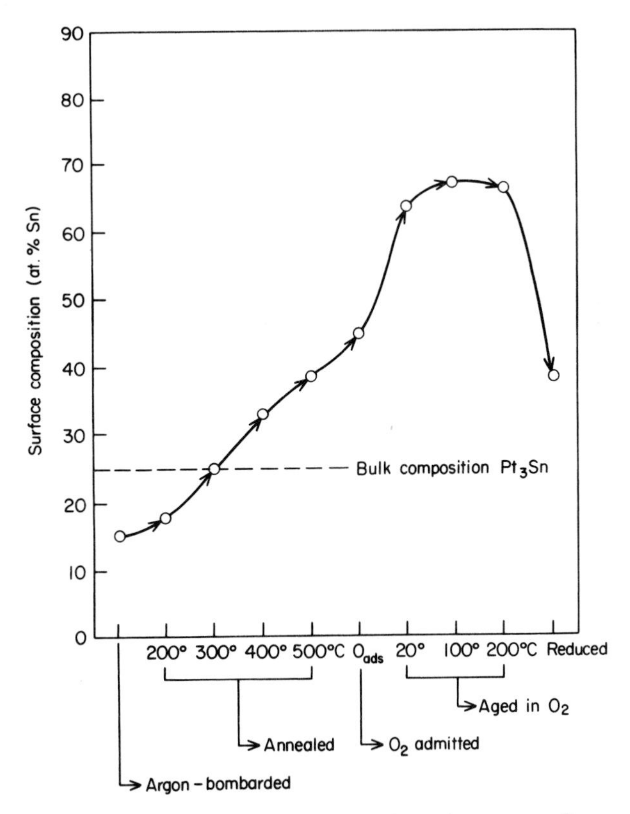

Fig. 12. Change of surface composition of Pt_3Sn throughout annealing and chemisorption.[11]

quantitatively analysed. Figure 12 demonstrates the change of surface composition of Pt_3Sn throughout annealing and chemisorption treatments studied by AES techniques.[11]

In an AES spectrum, especially a high resolution one, the shape and the position of the AES peaks reflect on the environment of the atom, in particular when the valence electrons are involved in the Auger process. Carbon deposited on a Mo surface and the carbon in chemisorbed CO, for example,

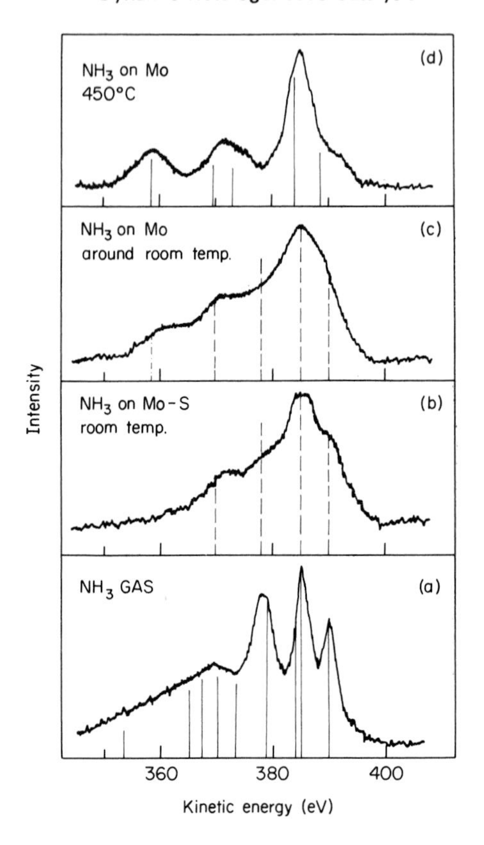

Fig. 13. High resolution, nonderivative nitrogen K–LL Auger spectra of (a) gaseous NH_3, (b) adsorbed NH_3 on a sulphur-segregated surface at room temperature, (c) on a clean surface at about room temperature, and (d) at 450°C. The vertical solid lines in (a) and (d) represent the energies calculated in the equivalent core approximation. The vertical broken lines in (b) and (c) represent observed energies and intensities for the peaks of gaseous NH_3. The dot–dashed line in (c) shows the observed value of Mo_2N which is at the lowest energy in (d).[12]

give different peak shapes, providing important information on the behaviour of surface species. High-resolution spectra of the Auger electrons from nitrogen in adsorbed species are given in Fig. 13.[12] The peaks in the spectra can be assigned on the basis of the "equivalent core approximation". When ammonia is introduced onto a Mo surface at 450°C, nitride formation can be demonstrated as shown in the figure, whereas ammonia adsorption on Mo at room temperature exhibits a decrease in NH bonding which suggests dissociative adsorption.† Auger electron spectroscopy may be readily combined with LEED to provide more information on surface phenomena.

2-5-3 *Photoelectron spectroscopic methods*

On irradiation by soft X-rays, for example Mg K_α (1254 eV) or Al K_α(1487 eV), atoms can be ionized by the emission of an electron from an inner shell (Fig. 14). If the energy of the X-ray photons is kept constant, the energy distribution

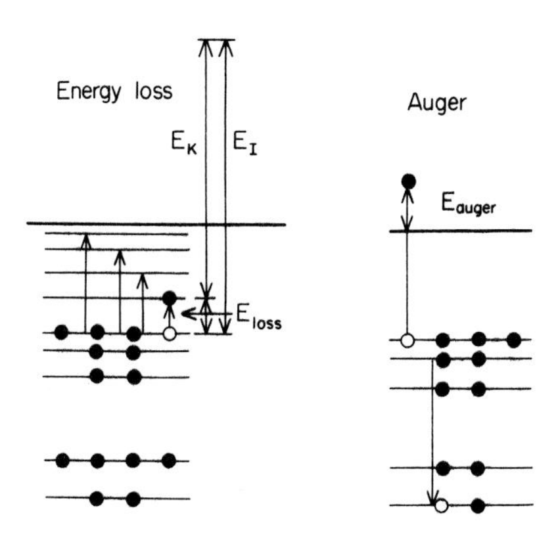

Fig. 14. Diagram of electron transfer.

of the emitted electrons provides information on the band structure of solid and also the energy levels of the electrons in the surface atoms, and this is characteristic of the types of atoms in the specimen. Consequently, as in AES, the photoelectron energy distribution may be used to provide quantitative and qualitative analyses of the surface species. Different environments of the atoms being studied give rise to different peak shapes and positions and these differences provide information on the nature of surface bonds and the state of the surface species. An example of the effect of chemical environment is given in Fig. 15[13] where the 1-s peaks of the carbons in methyl, methylene and carbonyl groups appear with different binding energies.

The surface of a Fe–V_2O_4 catalyst was studied by this X-ray photoelectron spectroscopy (XPS) before and after cyclohexane dehydrogenation.[14] The changes in the O(1s) and V(3s, 3p) peaks shown in Fig. 16 were interpreted as an electron shift from V atoms to O atoms.

† The intensity of 379 eV peak associated with the *K*(1e)(1e) transition decreased considerably. Since the 1e electronic state corresponds to the N—H σ-type bonding orbital, the decrease of this peak suggests that the N—H bond is dissociated at the chemisorption.

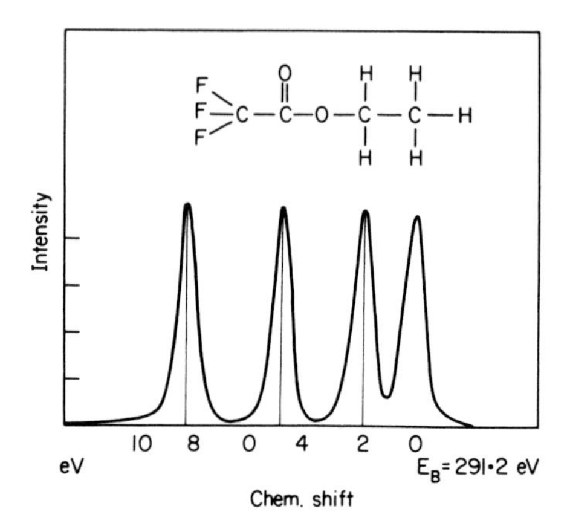

Fig. 15. Cls electron spectrum of the vapour of ethyltrifluoroacetate excited by monochromatized X-radiation. The four C lines follow in the same sequence as the structural formula in the figure.[13]

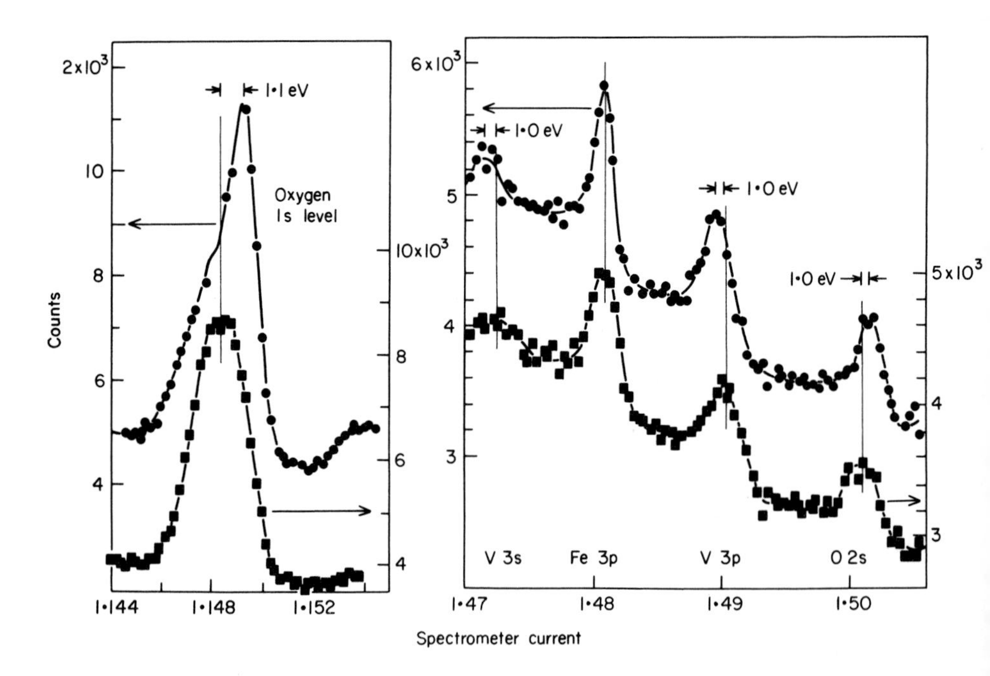

Fig. 16. Partial photoelectron spectra from FeV_2O_4 before (■) and after (●) use as a catalyst for the dehydrogenation of cyclohexane at 425°C.[14]

Instead of soft X-rays, lower energy photons can also be used to study photoemission energy distributions. Demuth and Eastman studied the photoemission from valence orbital energy levels of chemisorbed hydrocarbons and obtained a clear picture of the chemisorbed states of the adsorbed species. They used a photon energy of 21·2 eV from a He-resonance lamp equipped with a retarding-field analyser and measured the ionization energies of acetylene, ethylene and benzene chemisorbed on Ni (111) surface. The results, shown in Fig. 17, demonstrated π-orbital bonding shifts as well as non-chemical-bonding relaxation shifts for both π and σ orbitals.[15]

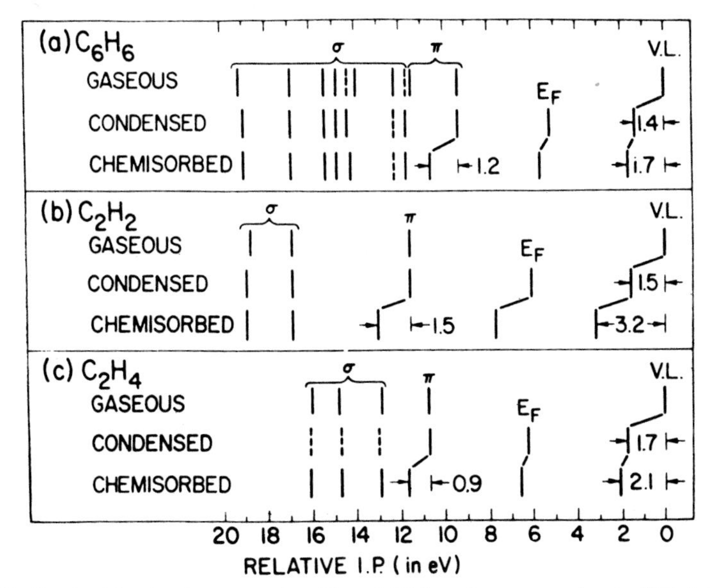

Fig. 17. Vertical ionization energies, Fermi (E_F) and vacuum levels (V.L.) for the gaseous, condensed and chemisorbed phases of (a) benzene (b) acetylene and (c) ethylene, all plotted relative to σ-orbital gas phase levels. Relaxation shifts are given for relevant π-orbital shifts. The dotted levels represent less certain orbital ionization energies.[15]

The dehydrogenation of chemisorbed ethylene to acetylene on Ni(111) surface can be directly observed from Fig. 18, since the spectrum of chemisorbed acetylene is similar to that for dehydrogenated ethylene. The π-d bonding interaction strengths and chemisorption energies were also estimated from the spectroscopic energy levels.

Other electron spectroscopic techniques such as ion neutralization spectroscopy (INS), Penning ionization electron spectroscopy and appearance potential spectroscopy (APS) will become more widely developed as tools for studying surface behaviour.

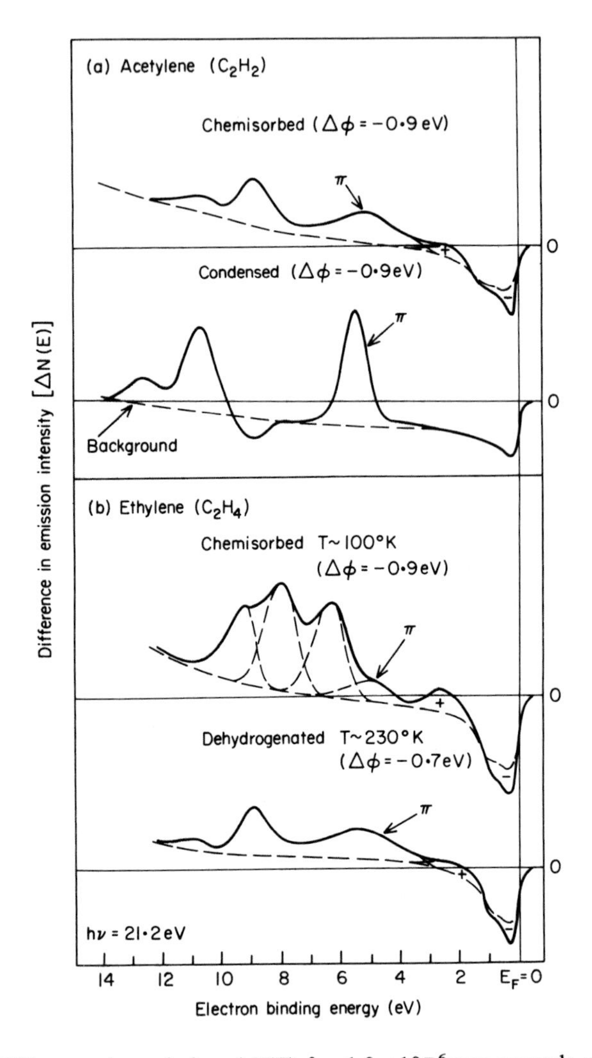

Fig. 18. (a) Difference in emission $\Delta N(E)$ for $1{\cdot}2\times10^{-6}$ torr-seconds exposure to acetylene at $T = 300\,\mathrm{K}$ (or at $T = 100\,\mathrm{K}$) and for condensed acetylene formed at $T = 100\,\mathrm{K}$ with acetylene pressures of 6×10^{-8} torr. (b) $\Delta N(E)$ for chemisorbed ethylene (exposure of $1{\cdot}2\times10^{-6}$ torr-seconds at $T = 100\,\mathrm{K}$ and for dehydrogenated ethylene (obtained by warming to $T = 230\,\mathrm{K}$ or with an initial exposure at $T = 300\,\mathrm{K}$).[15]

2-6 Spectrophotometric methods

Adsorbed species on catalyst surfaces may be studied directly by spectrophotometric methods such as infrared and ultraviolet spectroscopy. Infrared investigations have been widely used to study simple molecules such as carbon monoxide, and methanol in the adsorbed state, while ultraviolet spectroscopy has been applied to the study of electronic transitions in such species as carbenium ions and ion radicals.

Since Eischens and his co-workers studied the infrared spectra of carbon monoxide chemisorbed on metal surfaces, a great number of reports have appeared on the infrared absorption spectra of chemisorbed species.[16] The infrared technique is not always as sensitive as many other spectroscopic techniques, but one of its big advantages is that we can examine not only the structure of the chemisorbed species, but also follow their dynamic behaviour by means of isotopic substitution (See Sections 1-4, 4-1).

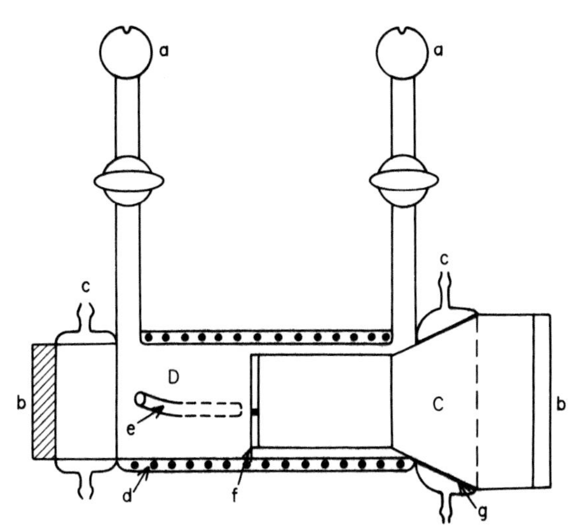

Fig. 19. Infrared cell. (C) Sample holder, (D) main body. (a) Ball joint, (b) sodium chloride windows, (c) cooling jackets, (d) heater, (e) thermocouple pocket, (f) glass rings for holding catalyst disk, (g) taper ground glass joint.

One type of infrared cell employed is shown in Fig. 19. The sample disk in the middle may be heated to 450°C by the electric heater and the two infrared transparent windows allow the absorption by the sample to be studied. For a system to be amenable to investigation by this technique, the solid sample should be as transparent as possible in the infrared region and the adsorbed species should have high extinction coefficients.

Absorption spectra of carbon monoxide adsorbed on metal surfaces exhibit two peaks; the one near 2000 cm^{-1} is assigned to M—C≡O and that round 1900 cm^{-1} to

```
M
  \
   C≡O
  /
M
```

(bridge structure). As shown in Fig. 20, the spectrum depends upon the metal on which the carbon monoxide is adsorbed. The formation of methane by the reaction of carbon monoxide with hydrogen is slow over copper and platinum, but fast over nickel and palladium. This activity for methane formation correlates with the characteristic absorption spectra of CO adsorbed on those metals.

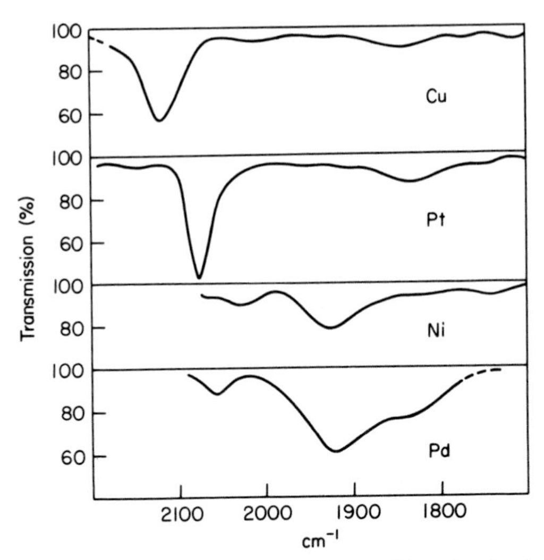

Fig. 20. Infrared absorption spectra of carbon monoxide adsorbed on various metal surfaces.

Hydrogen adsorbed on platinum exhibits two types of adsorption presumably Pt—H and

```
    H       H
   / \     / \
Pt     Pt     Pt,
```

while on zinc oxide adsorbed hydrogen gives rise to Zn—H and O—H as shown in Fig. 21. It is also possible to study vibrational spectra of certain species adsorbed on single crystals, as is now being done by Pritchard.[17]

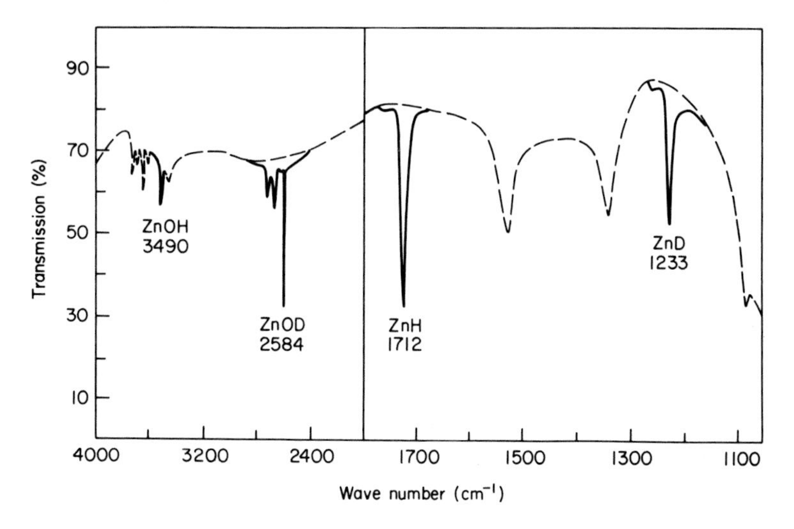

Fig. 21. Infrared absorption spectra of hydrogen chemisorbed on zinc oxide.

Recently a high sensitivity high-resolution spectrometer the fourier-transform spectrometer, has been developed. Smaller amounts of adsorbed species may be detected and the behaviour of adsorbed species may be followed in a more quantitative fashion.

The ultraviolet absorption spectra of chemisorbed species are also used to identify surface species, in many cases by comparing with the spectra of identified species. For example, triphenyl methane adsorbed on silica alumina exhibits the formation of a carbenium ion as given in Fig. 22.[18]

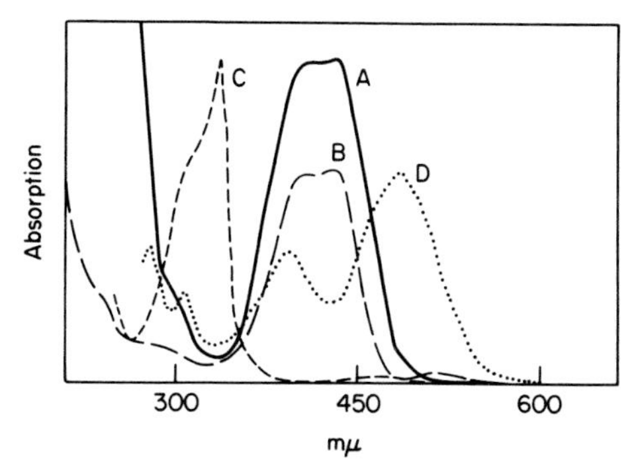

Fig. 22. Electronic spectra of (A) triphenylmethane adsorbed on silica–alumina, (B) triphenylmethyl alcohol in concentrated sulphuric acid, (C) triphenylmethyl radical, and (D) triphenyl methide ion.[18]

2-7 NMR and esr techniques

Nuclei with odd mass numbers such as protons, and species with unpaired electrons, possess an intrinsic spin angular momentum and a magnetic moment, and may be studied by NMR and esr techniques respectively.(19) In some cases, protons on solid surfaces have been studied by NMR techniques and adsorbed species such as hydroxyls, alcohols, acids, ketones and hydrocarbons on silica gel, alumina or other oxides such as magnesium oxide, titania and vanadium oxide have been investigated.

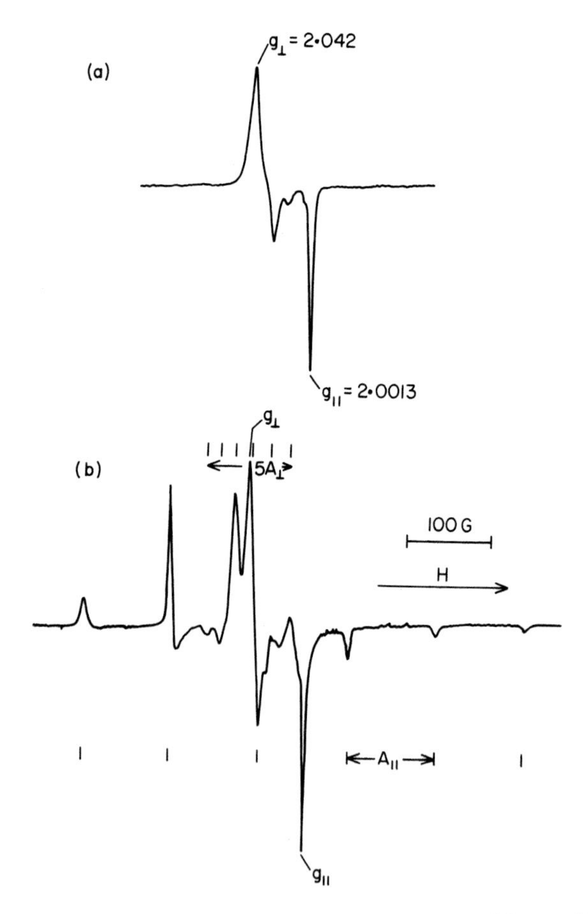

Fig. 23. (a) Spectrum of $^{16}O^-$ on magnesium oxide; (b) spectrum of O^- enriched with 51 % ^{17}O.(20)

Esr techniques are used to study various adsorbed species with unpaired electrons. When oxygen is adsorbed on oxides, three forms of oxygen, O^-, O_2^- and O_3^- have been observed by this technique.[20] When oxygen is adsorbed, for instance, on magnesium oxide with trapped electrons on the surface, an O^- spectrum with $g_\perp = 2{\cdot}042$ and $g_\parallel = 2{\cdot}0013$ is immediately obtained, as shown in Fig. 23. It can be confirmed that the spectrum is due to a monoatomic form of oxygen by using ^{17}O. Naccache demonstrated that the O^- species reacts readily with CO and ethylene to form CO_2^- and $C_2H_4O^-$, respectively, the species being identified by their esr spectra.[21]

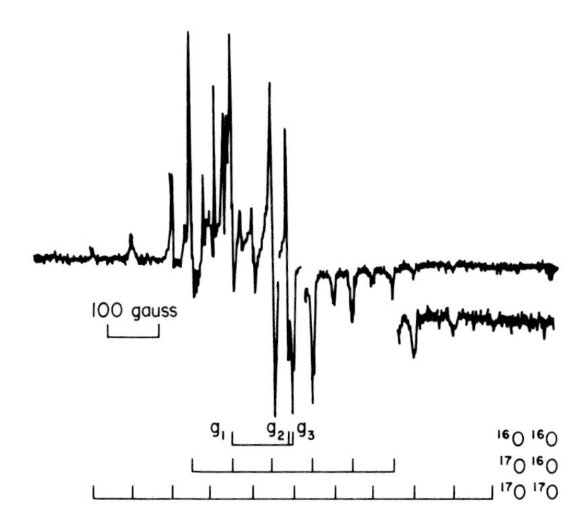

Fig. 24. The esr spectrum of O_2^- on MgO. The field increases from left to right and the gain has been reduced by five times for the central lines; a small portion of the high-field spectrum was overmodulated to show the outermost lines. For clarity, no attempt has been made in this diagram to insert levels for the lines around g_1.[20, 23]

Formation of the superoxide ion has also been widely studied. The nuclear spin of ^{17}O is 5/2 and, accordingly, a paramagnetic species containing one ^{17}O atom would be expected to exhibit six $(2I + 1)$ lines of the same amplitude for each principal direction. A diatomic $^{17}O^{17}O$ species would show eleven, $2(2I) + 1$, lines with amplitudes 1 : 2 : 3 : 4 : 5 : 6 : 5 : 4 : 3 : 2 : 1 for each principal direction. Wong and Lunsford have confirmed the ^{17}O hyperfine splitting for O_2^- on MgO (Fig. 24).

An ozonide ion is also formed from O^- and O_2, as has been demonstrated by Wong and Lunsford and shown in Fig. 25.[22]

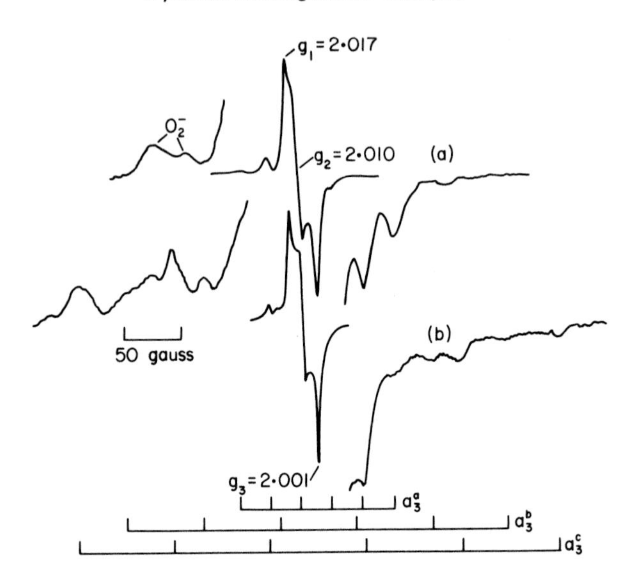

Fig. 25. (a) Spectrum of O_3^- produced by the reaction of $^{17}O^-$ (71·9% enrichment) with $^{16}O_2$; (b) spectrum of O_3^- produced by the reaction of $^{16}O^-$ with $^{16}O^{17}O$ (44·5% enrichment). The low- and high-field regions of the spectra were recorded at an increased power level and overmodulation in order to show the weak lines.[22]

2-8 Other techniques for studying adsorption

There are a variety of other techniques for studying adsorption, work function measurements (FEM is an example), calorimetry, ellipsometry, magnetic susceptibility and Mössbauer effect[24] and so forth. In some cases, flash desorption techniques or temperature programmed desorption can be used to study the strengths of adsorption of different species.

A typical desorption spectrum of the oxygen adsorbed on zinc oxide after exposure to 8 cm oxygen pressure at room temperature, produced, as shown in Fig. 26, two peaks with maxima at 180–190 and 285–295°C. An esr study revealed that the low-temperature peak corresponds to O_2^- and the high-temperature one is probably due to O^-. When a small pulse of carbon monoxide was passed over an oxygen preadsorbed zinc oxide surface at room temperature, the high-temperature peak disappeared, as shown in the figure. It was accordingly demonstrated that O^- is the reactive intermediate in the catalytic oxidation of carbon monoxide over zinc oxide and also that O^- does not change to O_2^- over zinc oxide at temperatures below 200°C.[25]

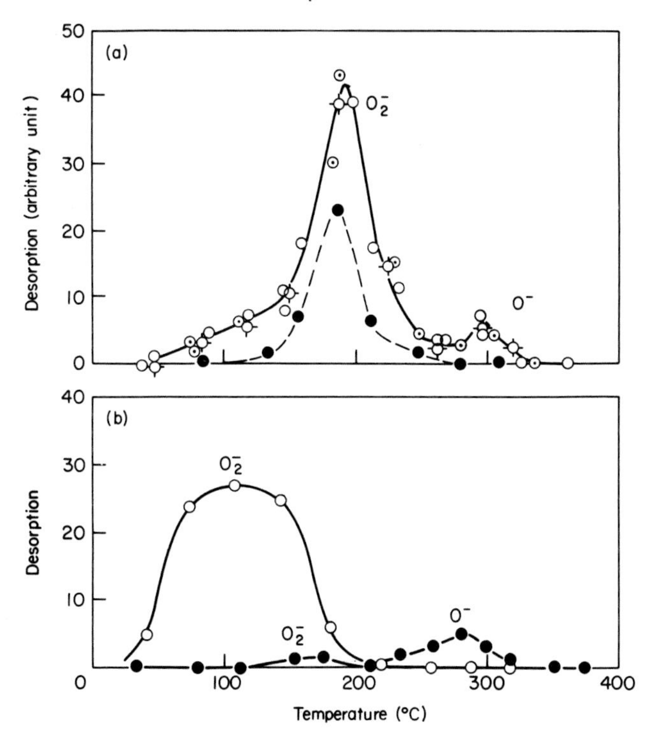

Fig. 26. Desorption spectra of oxygen from zinc oxide: (a) solid line represents desorption of oxygen adsorbed on evacuated ZnO and broken line represents desorption of oxygen after CO pulse has been passed through; (b) ○ represents desorption of oxygen adsorbed on hydrogen presorbed ZnO and ● represents desorption of oxygen adsorbed at 200°C and cooled to room temperature.[25]

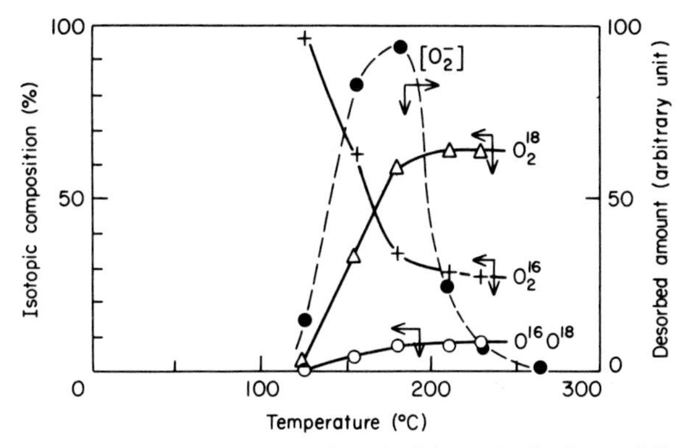

Fig. 27. Isotopic composition of oxygen desorbed from ZnO. Dotted line indicates approximate desorption.[26]

In a similar manner, Tanaka and Miyahara[26] studied the isotopic composition of oxygen desorbed from ZnO where $^{18}O_2$ had been adsorbed followed by $^{16}O_2$ adsorption at room temperature (Fig. 27). The isotopic composition of the desorbed oxygen changed with increasing temperature, and approached a uniform composition without isotopic mixing. That is, no $^{16}O^{18}O$ molecules appeared in this case except those originally contained in $^{18}O_2$. Thus, O_2^- does not take part in this oxygen exchange reaction. Furthermore the strength of the oxygen adsorption is variable because the $^{18}O_2$ that was adsorbed first desorbed later than the $^{16}O_2$. Similar results obtained by the addition of CO to ZnO on which oxygen ($^{18}O_2$) had been preadsorbed, revealed that O_2^- on ZnO is inactive for the oxidation of carbon monoxide and does not participate in isotopic exchange between oxygen and CO or CO_2 (Fig. 28).

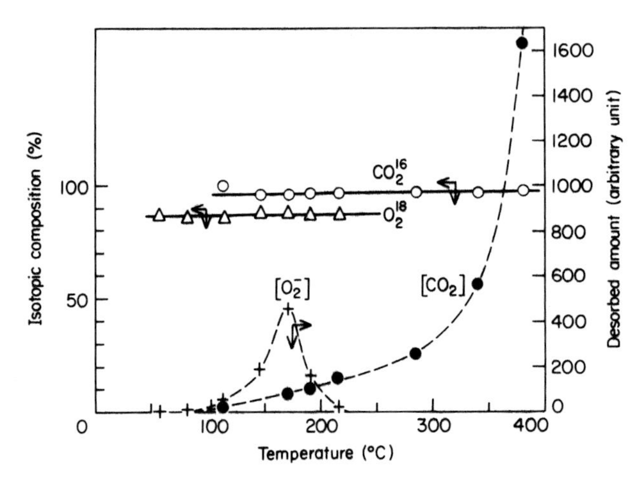

Fig. 28. Isotopic composition of desorbed oxygen and desorbed CO_2 when CO_2 was admitted on $^{18}O_2$ adsorbed ZnO.[26]

2-9 The nature of adsorbent–adsorbate bonds

Of the various techniques for studying adsorption that we have mentioned, not one alone is generally sufficient to identify the adsorbed state unequivocally. However, by focusing the light of various viewpoints and techniques on the problem, the real adsorbed state may be determined and its reaction mechanism elucidated.

In this section the nature of chemical bonds between adsorbent and adsorbate will be briefly outlined. The nature of adsorbent–adsorbate bonds is, overall, similar to that of bonds in general. For example, if ammonia is introduced onto Lewis acid sites on a surface, such as aluminium ions in alumina, adsorption takes place as an acid–base interaction. If, on the other hand, the ammonia adsorption occurs on Brönsted sites such as are on a *p*-toluene sulphonic acid crystal, or a surface which contains protons that can react with basic species, NH_4^+ will be formed. Olefins and aromatic compounds react with such Brönsted sites to form carbonium ions, and isomerization can take place in addition with olefins.

As shown in Fig. 21,[27] when hydrogen is adsorbed on ZnO, the hydrogen molecule heterolytically dissociates, and combines with the oxygen and zinc ions. A more clear-cut example of heterolytic dissociation is observed when hydrogen is adsorbed on electron–donor acceptor complexes such as the anthracene sodium complex, $An^{2-}2Na^+$. When hydrogen is introduced over the stoichiometric complex, photospectroscopic and NMR techniques indicate that the following reaction takes place in which H^+ and H^- ions

$$[\text{anthracene}]^{2-}\,2Na^+ + H_2 \rightleftharpoons [\text{9-H,H-anthracene}]^{-}\,Na^+\ NaH$$

combine with an anthracene anion and sodium ion to form a mononegative monohydroanthracene anion and sodium hydride, respectively.[28] The electron donor–acceptor interaction plays an important role in many adsorption systems; for instance, oxygen is adsorbed on zinc or nickel oxide by accepting an electron or electrons from the oxide as demonstrated in Section 2-7.

The formation of covalent surface complexes are often proposed in various adsorption systems. The adsorption of carbon monoxide and olefins on transition metals are frequently quoted as examples. The structures of the surface complexes of these gases are generally similar to those of metal carbonyls and metal–olefin complexes (Figs 20 and 7, respectively), which are already well known in the field of coordination chemistry. Such comparisons are well supported, for example, by observations from infrared and photoelectron spectroscopy, which are explained in Figs 20 and 17, respectively.

Pauling's approximation has also been adapted to the adsorption of hydrogen on metals and the metal–hydrogen bond strength is taken as the arithmetic average of the metal–metal and hydrogen molecule bond strengths. The difference in electronegativities between the metal and the hydrogen is also taken into consideration, giving the bond a small amount of ionic character. Such partially ionic surface complexes have been studied by work function changes and the creation of dipole moment when hydrogen is chemisorbed on metals.

A general rule in chemisorption was proposed by Tanaka and Tamaru.[29] According to the rule, the initial heats of chemisorption of gases on various metal surfaces are empirically expressed by the following equation:

$$Q_0 = \mathbf{a}[(-\Delta H_0^\circ) + 37] + 20 \text{ kcal/mol}$$

as shown in Fig. 29, where Q_0 is the initial heat of chemisorption, **a**, a constant which depends upon the electronegativity of gases adsorbed (as given in Fig. 30), and $-\Delta H_0^\circ$, the heat of formation of the highest oxide per metal atom.

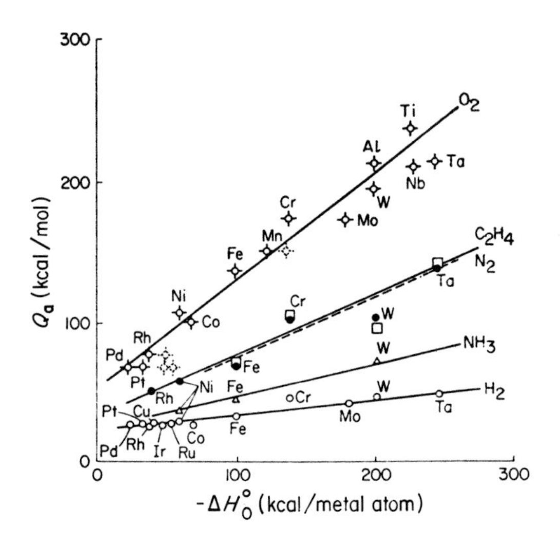

Fig. 29. The correlation between the initial heat of adsorption of various gases (Q_a) and the standard heat of formation of the highest oxide per metal atom ($-\Delta H_0^\circ$).[29]

Accordingly, the initial heats of chemisorption of a gas on various metal surfaces may be approximately estimated from its electronegativity (or **a** value) and $-\Delta H_0^\circ$. Such a general rule strongly suggests that the binding between adsorbate and adsorbent is chemical in nature, which is in accord with the general behaviour of chemical bonds.

It is anticipated that, as more and more experimental results are accumulated by means of various techniques, especially by modern physical techniques as described above, the theoretical aspects of the subject will certainly be advanced.

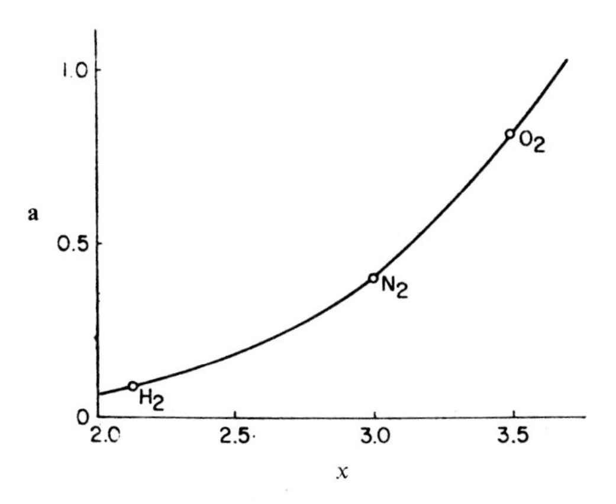

Fig. 30. The values of **a** for oxygen, nitrogen, and hydrogen are plotted against their electronegativity (x).[29]

2-10 Adsorption isotherm

When a gas is admitted onto a surface, the amount of adsorption increases with time finally reaching equilibrium with the ambient gas. If we keep the temperature constant but decrease the ambient pressure, the amount adsorbed decreases correspondingly. At a higher temperature under the same pressure, the amount adsorbed is usually less, because adsorption is generally an exothermic process. The relation between the amount of gas adsorbed and its equilibrium pressure is called an adsorption isotherm. Many types of adsorption isotherm have been proposed but one of the most familiar is the Langmuir adsorption isotherm.

The rate of adsorption (r_a) is dependent upon the pressure (P) of the gas to be adsorbed and the number of vacant sites on the surface. The rate of desorption (r_d) is a function of the fraction (θ) of the sites on the surface occupied by the adsorbed species. If it is assumed that adsorption takes place when an adsorbate molecule collides with a vacant surface site, the rate of adsorption is proportional to the gas pressure and to the number of vacant sites.

$$r_a = k_a P(1 - \theta) = k_a^\circ \exp(-E_a/RT)P(1 - \theta)$$

Here it is assumed that k_a and k_a° are constants and that k_a is independent of θ. In this equation, E_a is the activation energy for the adsorption.

In a similar manner, if the rate of desorption is proportional to the surface coverage, θ,

$$r_d = k_d\theta = k_d^\circ \exp(-E_d/RT)\theta$$

where E_d is the activation energy for desorption. At equilibrium, r_a is equal to r_d. Consequently,

$$k_a^\circ \exp(-E_a/RT)P(1-\theta) = k_d^\circ \exp(-E_d/RT)\theta$$

$$\frac{\theta}{(1-\theta)} = (k_a^\circ/k_d^\circ)\exp[(E_d - E_a)/RT]P$$

Since $(E_d - E_a)$ is equal to the heat of adsorption (Q_{ad}),

$$\frac{\theta}{(1-\theta)} = (k_a^\circ/k_d^\circ)\exp(Q_{ad}/RT)P = bP$$

where b corresponds to an equilibrium constant and is equal to $(k_a^\circ/k_d^\circ)\exp(Q_{ad}/RT)$. Consequently,

$$\theta = bP/(1 + bP)$$

This is the equation for the well known Langmuir adsorption isotherm. Figure 31 illustrates a Langmuir adsorption isotherm using, as an example, hydrogen sorption by the graphite–alkali metal complex $C_{24}Rb$.[30]

If two gases, A and B, are adsorbed simultaneously and competitively on a surface, their surface coverages may be obtained in a similar manner:

$$r_a^A = k_a^A P_A(1 - \theta_A - \theta_B)$$

$$r_a^B = k_a^B P_B(1 - \theta_A - \theta_B)$$

$$r_d^A = k_d^A \theta_A$$

$$r_d^B = k_d^B \theta_B$$

where P_A and P_B are the partial pressure of A and B, respectively and k_a^A, k_a^B, k_d^A and k_d^B are constants assumed to be independent of the surface coverages θ_A and θ_B.

At equilibrium, the rates of adsorption and desorption are equal and we may consequently derive the following equations:

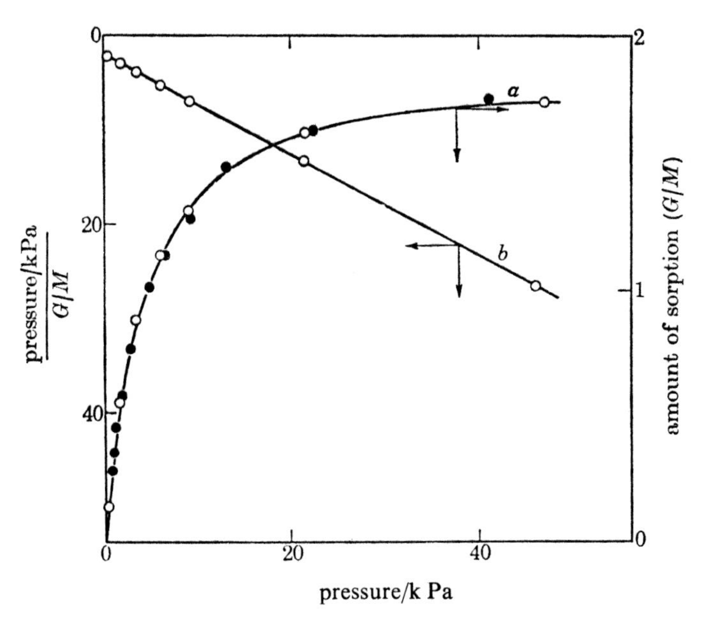

Fig. 31. Sorption–desorption curves for the system $C_{24}Rb$–D_2 at 113 K. (a) Pressure–sorption curves; ○, sorption; ●, desorption. (b) Langmuir plots.[30]

$$\theta_A = b_A P_A/(1 + b_A P_A + b_B P_B)$$

$$\theta_B = b_B P_B/(1 + b_A P_A + b_B P_B)$$

where b_A and b_B are $(k_a^{\circ A}/k_d^{\circ A})\exp(Q_{ad}^A/RT)$ and $(k_a^{\circ B}/k_d^{\circ B})\exp(Q_{ad}^B/RT)$ respectively. If more than two gases are adsorbed at the same time, the following adsorption isotherms are similarly obtained:

$$\theta_A = b_A P_A\Big/\left(1 + \sum_i b_i P_i\right)$$

$$\theta_B = b_B P_B\Big/\left(1 + \sum_i b_i P_i\right)$$

According to this adsorption isotherm, the adsorption is proportional to the pressure at lower pressures where $1 \gg \sum_i b_i P_i$, but becomes independent of pressure at higher pressures of A when $b_A P_A \gg 1 + \sum_j b_j P_j$ and the surface is saturated with A.

In the derivation of these adsorption isotherms it is assumed that (1) no surface complexes are formed between the A and B species, such as, for example, carbonate formation between CO_2 and O_2, or formate between CO_2 and H_2, (2) the heat of adsorption and the activation energies for adsorption and desorption are independent of the surface coverage, (3) multilayer adsorption does not take place and the rate of adsorption is proportional to the fraction of vacant sites and to the partial pressure of the adsorbing gas, while the desorption rate is proportional to the surface coverage of the species under consideration.

If we extend the treatment of Langmuir isotherm to multilayer adsorption, we may get the "BET" adsorption isotherm[31]

$$\frac{P}{v(P_0 - P)} = \frac{1}{v_m C} + \frac{C - 1}{v_m C}\frac{P}{P_0}$$

where P is the pressure of the gas adsorbed; P_0, its saturated vapour pressure; v, the amount adsorbed under P; C, a constant; and v_m, the amount of gas required for monolayer formation. This adsorption isotherm may be employed to estimate the surface area of the adsorbent from the value of v_m and the size of the gas molecule adsorbed.

If the adsorption is dissociative and requires two adjacent vacant sites, the rates of adsorption and desorption and the adsorption isotherm are given as follows (Chapter 3, Fig. 4)

$$r'_a = k'_a P(1 - \theta)^2$$

$$r'_d = k'_d \theta^2$$

$$\theta/(1 - \theta) = (b'P)^{\frac{1}{2}}$$

or

$$\theta = \sqrt{b'P}/(1 + \sqrt{b'P})$$

In the derivation of the Langmuir adsorption isotherm it is assumed that the heat of adsorption is independent of surface coverage. In many adsorption systems the heat of adsorption may be measured by indirect thermodynamic methods or by a direct calorimetric technique. In the former method, adsorption equilibria are measured at various temperatures as shown in Chapter 3, Fig. 3, and the gas pressures corresponding to a fixed surface coverage are obtained. The heat of adsorption may then be determined from the Clapeyron–Clausius equation;

$$(\mathrm{d}(\ln P)/\mathrm{d}T)_\theta = \Delta H/RT^2$$

where ΔH, (equal to $-Q_{ad}$), is the increase in enthalpy due to adsorption when the surface coverage is θ. The heats of adsorption may be determined experimentally at various coverages and these are called differential or isosteric heats of adsorption. For many chemisorption systems, the heat of adsorption obtained experimentally generally decreases with surface coverage as shown in Fig. 32.[32]

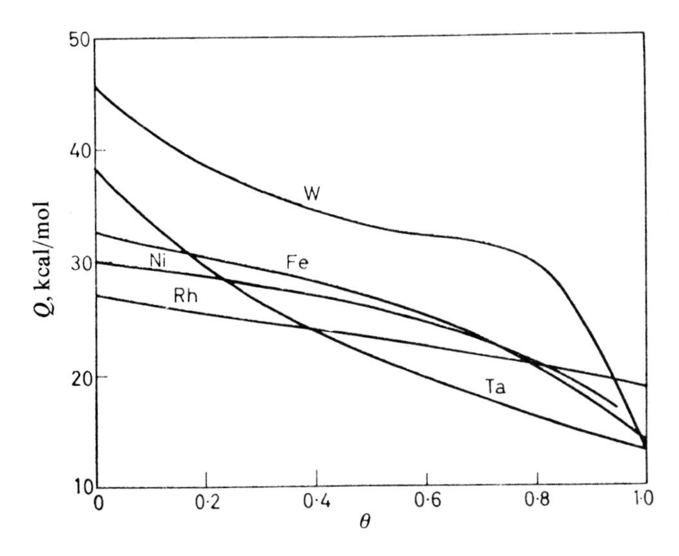

Figure 32. The dependence of the heat of adsorption of hydrogen upon its coverage.[32]

If we assume that the heat of adsorption decreases linearly with coverage, we obtain an adsorption isotherm:

$$\frac{\theta}{1-\theta} = b_0 \exp((Q_{ad} - a\theta)/RT)P$$

or

$$A(\ln(1-\theta)/\theta + \ln BP) = \theta$$

where A and B are constant at constant temperature and Q_{ad} is the differential heat of adsorption when $\theta = 0$. If we neglect $\ln(1 - \theta)/\theta$ as, at moderate surface coverages, its value is small and relatively independent of θ, the following form of adsorption isotherm is obtained.

$$\theta = A \ln BP$$

This is often called the Frumkin–Temkin isotherm.[33] An example of an

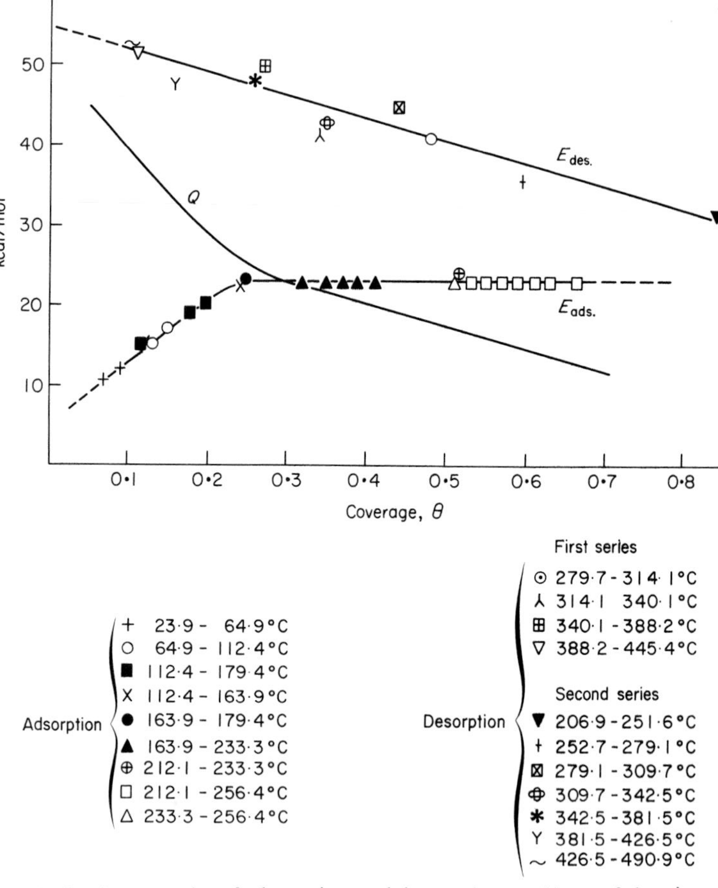

Fig. 33. Activation energies of adsorption and desorption, and heat of chemisorption as functions of coverage, for nitrogen on a singly promoted, intensively reduced iron catalyst.[34]

E_{des} = activation energy of desorption
E_{ads} = activation energy of adsorption
$Q = E_{des} - E_{ads}$ (heat of adsorption).

adsorption system which behaves in accordance with this adsorption isotherm is shown in Fig. 33.[34]

If the heat of adsorption decreases with coverage as $Q_{ad} - a \ln \theta$, the following adsorption isotherm may be obtained in a similar manner:

$$\frac{\theta}{(1-\theta)} = b_0 P \exp\big((Q_{ad} - a \ln \theta)/RT\big)$$

$$\theta = A' P^{1/n}$$

where A' and n are constants and usually $1 < n < 10$. This adsorption isotherm is sometimes called the Freundlich isotherm, and an example is

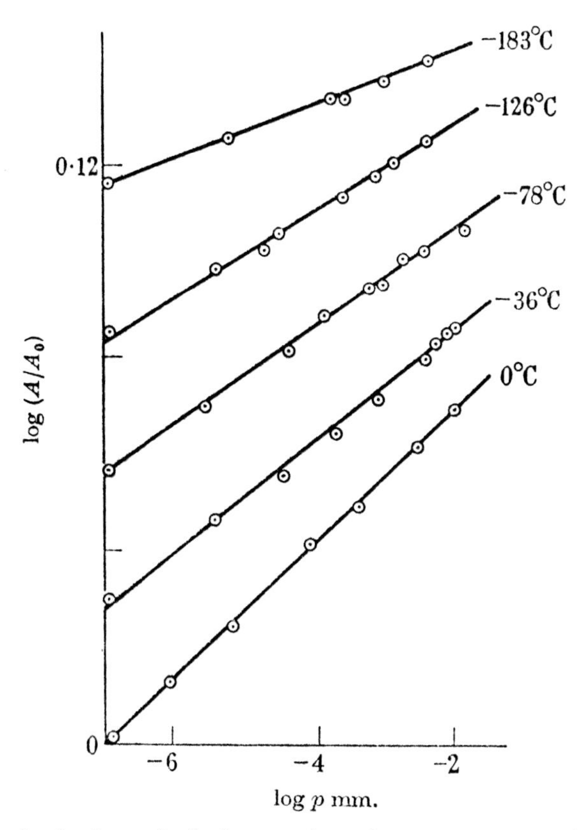

Fig. 34. Log–log isotherms for hydrogen adsorption on an evaporated tungsten film, where A is the adsorption, A_0 the adsorption irreversible to pumping at 0°C and p the pressure.[35]

shown in Fig. 34. The heat of adsorption of hydrogen on tungsten, demonstrated in Fig. 34, changes with coverage as follows:

$\theta(\%)$	80	85	90	95	100
Isothermal heat (kcal)	15·2	12·4	9·3	6·1	3·0

The heat of adsorption may decrease with surface coverage for various reasons. Adsorption starts at the stronger adsorption sites on the surface. The properties of a homogeneous metal surface are influenced by the presence of adsorbed species such that the work function (or electron affinity or ionization potential) may be decreased (or increased) by adsorption and it is subsequently more difficult for further adsorbed species to transfer electrons to (or from) the surface, leading to weaker adsorption as the coverage increases. Repulsion between the adsorbed species may also cause the heat of adsorption to decrease with coverage. As the heat of adsorption decreases, the reactivity of the adsorbed species will also change and this is important in heterogeneous catalysis. In this connection it should be emphasized that we have to be always careful in applying homogeneous kinetics to heterogeneous reactions, and in deriving rate equations on a Langmuir model without knowing the departures of practical systems from the idealized ones.

References

1. G. C. Bond, "Catalysis by Metals", Academic Press, p. 66 (1962); D. O. Hayward and B. M. W. Trapnell, "Chemisorption", Butterworth (1964).
2. R. Gomer, "Field Emission and Field Ionization", Oxford Univ. Press (1961).
3. E. W. Müller, "Modern Res. Techniques in Phys. Metallurgy", Am. Soc. of Metals, Cleveland, p. 33 (1953).
4. Z. Knor and E. W. Müller, *Surf. Sci.* **10,** 21 (1968).
5. E. W. Müller, *Quart. Rev.* **23,** 177 (1969).
6. J. Block, *J. Vac. Sci. Tech.* **7,** 63 (1970).
7. T. N. Rhodin and D. S. Y. Tong, *Physics Today*, **28**, No. 10 (1975).
8. A. E. Morgan and G. A. Somorjai, *J. Chem. Phys.* **51**, 3309 (1969); see G. A. Somorjai, *Accnts. Chem. Res.* **9**, 248 (1976); *J. Catalysis*, **42**, 181 (1976).
9. F. M. Propst and T. C. Piper, *J. Vac. Sci. Tech.* **4,** 53 (1967).
10. T. Kawai, K. Kunimori, T. Kondow, T. Onishi and K. Tamaru, *J. Chem. Soc. Faraday I*, **70,** 137 (1974); Proc. 2nd Internl. Conf. on Solid Surf. 1974 Japan, *Appl. Phys.* **2,** 513 (1974); K. Kunimori, T. Kawai, T. Kondow, T. Onishi and K. Tamaru, *Surf. Sci.* **46,** 567 (1974).
11. R. Bouwman, L. H. Toneman and A. A. Holscher, *Surf. Sci.* **35,** 8 (1973).
12. T. Kawai, K. Kunimori, T. Kondow, T. Onishi and K. Tamaru, *Phys. Rev. Lett.* **33,** 533 (1974).
13. K. Siegbahn, *Endeavour*, **32,** 51 (1973).
14. W. N. Delgass, T. R. Hughes and C. S. Fadly, *Catalysis Rev.* **4,** 179 (1970).

15. D. E. Eastman and J. E. Demuth, Proc. 2nd Internl. Conf. on Solid Surf. 1974 Japan, *Appl. Phys. Suppl.* **2**, 827 (1974); *J. Vac. Sci. Technol.* **13**, 283 (1976).
16. R. P. Eischens and W. A. Pliskin, *Adv. in Catalysis*, **9,** 662 (1957); **10,** 1 (1958).
17. J. Pritchard and M. L. Sims, *Trans. Faraday Soc.* **66,** 427 (1970); M. A. Chesters, J. Pritchard and M. L. Sims, *Chem. Commun.* **1970,** 1454.
18. H. P. Leftin and M. C. Hobson, Jr., *Adv. in Catalysis*, **14,** 115 (1963).
19. E. G. Derouane, J. Fraissard, J. J. Fripiat and W. E. E. Stone, *Catalysis Rev.* **7,** 121 (1973).
20. J. H. Lunsford, *Catalysis Rev.* **8,** 135 (1973).
21. C. Naccache, *Chem. Phys. Lett.* **11,** 323 (1971).
22. N. B. Wong and J. H. Lunsford, *J. Chem. Phys.* **56,** 2664 (1972).
23. A. J. Tench and P. J. Holroyd, *Chem. Commun. 1968*, 471.
24. W. N. Delgass and M. Boudart, *Catalysis Rev.* **2,** 129 (1968).
25. K. Tanaka and G. Blyholder, *J. Phys. Chem.* **76,** 3184 (1972).
26. K. Tanaka and K. Miyahara, *Chem. Commun.* **1973,** 877.
27. S. Naito, H. Shimizu, E. Hagiwara, T. Onishi and K. Tamaru, *Trans. Faraday Soc.* **67,** 1519 (1971).
28. K. Tamaru, *Adv. in Catalysis*, **20,** 327 (1969); *Catalysis Rev.* **20,** 327 (1970).
29. K. Tanaka and K. Tamaru, *J. Catalysis*, **2,** 366 (1963).
30. K. Watanabe, T. Kondow, M. Soma, T. Onishi and K. Tamaru, *Proc. Roy. Soc.* (London), **A333,** 51 (1973).
31. S. Brunauer, P. H. Emmett and E. Teller, *J. Am. Chem. Soc.* **60,** 309 (1938).
32. O. Beeck, *Disc. Faraday Soc.* **8,** 118 (1950).
33. A. N. Frumkin and A. I. Shyrgin, *Acta Physiochim. USSR*, **3,** 791 (1935); M. I. Temkin and V. Pyzhev, *ibid.* **12,** 327 (1940).
34. J. J. F. Scholten and P. Zwietering, *Trans. Faraday Soc.* **53,** 1363 (1957); **55,** 2166 (1959).
35. B. M. W. Trapnell, *Proc. Roy. Soc.* (London), **A206,** 39 (1951).
36. H. Froitzheim, H. Ibach and S. Lehwald, *Phys. Rev. Lett.* **36**, 1549 (1976).
37. S. Donaich, I. Lindau, W. E. Spicer and H. Winick, *J. Vac. Sci. Technol.* **12,** 1123 (1975); T. Gufstafsson, E. W. Plummer, D. E. Eastman and J. L. Freeouf, *Solid State Commun.* **17**, 391 (1975); J. Anderson and G. J. Lapeyre, *Phys. Rev. Lett.* **36,** 376 (1976).
38. E. A. Stern, *J. Vac. Sci. Technol.* **14**, 461 (1977).

Appendix

The surface vibrations of H adsorbed on W(100) at 300 K have been studied by electron-energy-loss spectroscopy[36] and two vibration modes corresponding to atomic hydrogen in two different sites, on-top sites (at low coverages) and bridge sites (at high coverages), were assigned. Synchrotron radiation has been playing an increasingly important role in surface studies [37] and also in EXAFS (extended X-ray adsorption fine structure).[38]

Chapter 3

Kinetics of Catalytic Reactions on Solid Surfaces: Their Interpretation and the Elucidation of Reaction Mechanisms

3-1 The Langmuir rate law

A catalytic reaction involving one or more gaseous species reacting on a surface can be considered to take place via several stages: (1) diffusion of a reactant or reactants to the surface, (2) chemisorption of at least one of the reactants, (3) surface reaction between adsorbed species or between an adsorbed species and a colliding molecule from the ambient gas to form a product or products on the surface, (4) desorption of the product, and (5) diffusion away from the surface. Any of these steps, through which the overall reaction proceeds, may be rate-determining under certain reaction conditions.

Let us first consider the kinetics of catalytic reactions under simple conditions. As we have learned in equation (1-5-12), when a reactant A changes to a product or products D, the rate may be expressed by the following rate equation provided the constants k and k' are independent of the coverage of A and only A is adsorbed on the surface:

$$\mathrm{A} \rightarrow \mathrm{D}$$

$$-\frac{\mathrm{d(A)}}{\mathrm{d}t} = \frac{k\mathrm{(A)}}{1 + k'\mathrm{(A)}} \qquad (1\text{-}5\text{-}13)$$

The same rate expression may be obtained if we assume that the rate (R) is proportional to the amount of A adsorbed on the catalyst surface and that the Langmuir adsorption isotherm is applicable. This gives:

$$R = k\theta_{\mathrm{A}} = k\frac{b_{\mathrm{A}}p_{\mathrm{A}}}{1 + b_{\mathrm{A}}p_{\mathrm{A}}} \qquad (3\text{-}1\text{-}1)$$

According to equation (1-5-13) or (3-1-1), the rate of the reaction is first order with respect to A at low p_A or when A is weakly adsorbed, since $b_A p_A$ is much smaller than unity. When p_A is large and/or A is strongly adsorbed, $b_A p_A$ is much larger than unity, the catalyst surface is almost fully covered and the reaction is zero order with respect to A. When the catalyst surface is moderately covered by A, equations (1-5-13) and (3-1-1) predict that the rate will be fractional order with respect to A.

This has actually been found to be the case for some unimolecular catalytic decomposition reactions. For example, the decomposition of phosphine on a tungsten[1] or molybdenum[2] surface proceeds as a first order reaction at lower pressure, fractional order at moderate pressures and zero order at higher pressures, as shown in Table I.[1,2]

Table I. Decomposition of phosphine on tungsten and molybdenum catalysts (the activation energies are given in parentheses)[1,2]

	Temp.	$v = k(PH_3)$	$v = \dfrac{k(PH_3)}{1 + b(PH_3)}$	$v = k(PH_3)^{\circ}$
W	610–720°C	10^{-3}–10^{-2}	0·2	1–5 (Torr)
		(26·5)	(32·0)	(31·3 kcal/mol)
Mo	570–645°C	~0	0·06	0·20 (Torr)
		(15·1)	(20·8)	(22·3 kcal/mol)

If one of the products of the reaction is chemisorbed to an appreciable extent, it influences the form of the kinetic expression. In the decomposition of N_2O on Mn_3O_4, the following rate equation is obtained:[3]

$$R = \frac{k p_{N_2O}}{1 + b p_{N_2O} + b' p_{O_2}^{\frac{1}{2}}} \tag{3-1-2}$$

which is interpreted as indicating that oxygen is dissociatively adsorbed on the catalyst during the decomposition reaction, competing for the catalyst sites with the reactant N_2O.

This treatment may be extended to cover various possibilities. In a catalytic reaction between A and B, the rate may be assumed to be proportional to their coverages, θ_A and θ_B as follows:

$$R = k\theta_A\theta_B = \frac{k b_A b_B p_A p_B}{(1 + b_A p_A + b_B p_B)^2} \tag{3-1-3}$$

where no adsorption of the reaction products is taken into account.

Equation (3-1-3) can be simplified into various forms depending upon the relative sizes of $b_A p_A$ and $b_B p_B$: for example,

$$R = k' p_A p_B, \quad \text{if } b_A p_A,\ b_B p_B \ll 1 \tag{3-1-4}$$

$$R = k'' p_A / p_B, \quad \text{if } b_B p_B \gg b_A p_A, 1 \tag{3-1-5}$$

Equation (3-1-5) may be interpreted to mean that the catalyst surface is mainly covered by B, the "most-abundant surface species", to such an extent that it retards the overall reaction by excluding A from the catalyst surface. In other words, if one of the reactants, B, is strongly chemisorbed, the rate of reaction is first order to the partial pressure of B at low pressures, passes through a maximum and then decreases as the pressure of B is increased and saturates the catalyst sites. Such behaviour is illustrated in Fig. 1.

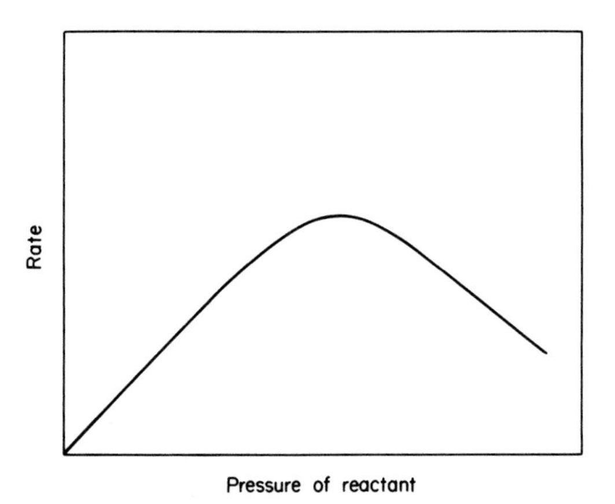

Fig. 1. Schematic representation of the variation of rate with the pressure of either reactant in the case of a bimolecular surface reaction proceeding by interaction between the two adsorbed molecules.

The apparent activation energy (E_{ap}) of a reaction is directly obtainable from the temperature-dependence of the apparent rate constant.

$$\mathrm{d} \ln k / \mathrm{d}T = E_{ap} / RT^2$$

For catalytic reactions on solid surfaces, the apparent activation energy often depends on the pressure range at which it is measured. For example, if rate equation (3-1-1) is applicable, at lower pressures of A the reaction is first order with respect to A and the apparent activation energy is equal to the activation energy (E) minus the heat of adsorption (Q_{ad}) contained in b_A

$$E_{ap} = E - Q_{ad}$$

as k and b_A can be expressed as $k_0 e^{-E/RT}$ and $b_0 e^{Q_{ad}/RT}$, respectively, where k_0 and b_0 are temperature independent. At higher temperatures the amount of A adsorbed at constant p_A decreases resulting in an apparent decrease in the "true" activation energy by an amount equal to the heat of adsorption.

However, at higher pressures of A, the catalyst sites are fully occupied, the coverage remaining constant at different temperatures, and the temperature-dependence of the rate is therefore equal to that of k, that is,

$$E_{ap} = E$$

In the case of a reaction described by equation (3-1-3), at lower pressures of B and A, the apparent activation energy will be the true activation energy E minus the sum of the heats of adsorption of A and B;

$$E_{ap} = E - Q_{ad}^{A} - Q_{ad}^{B}$$

and at high pressure of B where the overall reaction is retarded by B, the apparent rate constant would be kb_A/b_B and the apparent activation energy will be equal to $E - Q_{ad}^{A} + Q_{ad}^{B}$.

In the case of ethylene hydrogenation, both kinetic expressions (3-1-4) and (3-1-5) have been reported in the forms:

$$R = kP_E P_H \quad \text{(on Cu, 150–300°C)}^{(4)}$$

$$R = k'P_H/P_E \quad \text{(on Cu, 0–20°C)}^{(5)}$$

where P_E and P_H denote the partial pressures of ethylene and hydrogen, respectively.

The rate equation for this type of reaction is generally written in the form:

$$R = \frac{k\prod p_i^{\alpha_i}}{\left(1 + \sum b_i^m p_i^m\right)^n} \qquad (3\text{-}1\text{-}6)$$

where the exponents m and n are frequently unity, but may sometimes equal to $\frac{1}{2}$ and 2 for dissociative adsorption and bimolecular surface reaction, respectively. A rate law in the form of (3-1-6) is generally called a Langmuir rate law and this form was made popular by Hinshelwood, Schwab, Hougen, Watson and others.

The derivation of rate equations may be further extended to other possible reaction mechanisms, as for example when the surface reaction occurs when a gas molecule (or a very weakly adsorbed molecule) collides with an adsorbed species ("Eley–Rideal mechanism"). If the reaction takes place between A(g) and B(a), instead of being expressed by equation (3-1-3), the rate will be given by:

$$R = kp_A\theta_B = \frac{kb_B p_A p_B}{1 + b_A p_A + b_B p_B} \tag{3-1-7}$$

In a reaction, $A \rightarrow B + C$, if the reaction takes place through steps (1)–(4) as follows and step (3) is rate-determining, the overall forward rate may be expressed as equation (3-1-9).

$$\begin{aligned} &(1)\quad A(g) \rightleftarrows A(a) \\ &(2)\quad A(a) \rightleftarrows B(a) + C(a) \\ &(3)\quad B(a) \nrightarrow B(g) \\ &(4)\quad C(a) \rightleftarrows C(g) \end{aligned} \tag{3-1-8}$$

$$R = \frac{kb_B p_A/p_C}{1 + b_A p_A + b_B p_A/p_C + b_C p_C} \tag{3-1-9}$$

In this case B(a) is in quasi-equilibrium with A(g) and C(g), and not with B(g).† The rate expression (3-1-9) may take the form of equation (3-1-10) if the denominator is nearly unity, or if the catalyst surface remains almost unoccupied.

$$R = kp_A/p_C \tag{3-1-10}$$

If the second step of (3-1-8) is rate-determining, the rate will be expressed by the following equation:

$$R = k\theta_A = \frac{kb_A p_A}{1 + b_A p_A + b_B p_B + b_C p_C} \tag{3-1-11}$$

† $A(g) \rightleftarrows B(a) + C(g)$.

and this may also take the form of (3-1-10) provided $b_C p_C$ is much larger than the other terms in the denominator of (3-1-11), or if the surface sites are mainly occupied by C(a).

In the kinetic treatments that have been studied it has been shown that various kinetic expressions can be derived from different postulated mechanisms, and in some cases the same rate equation may be derived from different reaction mechanisms or from different assumptions.

In practice the rate of catalytic reactions is measured experimentally and the rate expression is obtained as a function of the partial pressures of the reactant and product gases. The next step is to interpret the rate equation to determine the reaction mechanism indicated by the kinetic behaviour. In this step, a reaction mechanism which may *explain* the experimental rate equation is often chosen without taking into account the many assumptions that have been made in the derivation of the theoretical rate equation. That a proposed mechanism explains an experimentally observed rate equation does not necessarily mean that it is true. Many assumptions employed in the derivation of the Langmuir rate law are not always true: it is generally accepted that the enthalpy of adsorption (or b values) and the reactivity of the adsorbed species (k values) are dependent on the surface coverage and that the catalyst sites are not uniform in their behaviour. Consequently, even if we calculate the heat of adsorption from the temperature-dependence of the b values, approximating the rate equation with the Langmuir rate laws, it does not necessarily demonstrate that the heat of adsorption is independent of the coverage. Without any experimental observations on the state of the adsorbed species under reaction conditions, discussions on reaction mechanisms can be continued indefinitely without any fruitful conclusions being reached. In some cases, the kinetics of adsorption and desorption may be obtained as functions of surface coverages in forms different from the Langmuir adsorption kinetics.

Let us take ammonia synthesis over a doubly promoted iron catalyst as an example. The rates of nitrogen chemisorption (r_a) and desorption (r_d) on the catalyst in the presence of hydrogen have been experimentally examined. They are expressed by the following equations as a function of nitrogen coverage, i.e. the fraction of the surface covered by the adsorbed nitrogen (θ_N), and nitrogen pressure, P_N.

$$\begin{aligned} r_a &= k_a P_N \exp(-g\theta_N) \\ r_d &= k_d \exp(h\theta_N) \end{aligned} \qquad (3\text{-}1\text{-}12)$$

where k_a, k_d, g and h are constant at constant temperature.

Suppose the nitrogen adsorption and desorption steps are rate-determining for ammonia synthesis and decomposition, respectively. The rate equation for the overall reaction (1-1-1) may be obtained as follows.

As the rate-determining step is the nitrogen adsorption or desorption, the chemical potential drop for this step will be essentially equal to that of the overall reaction, all the other steps being in equilibrium. Accordingly, the adsorbed nitrogen, $N_2(a)$, is not in equilibrium with the nitrogen gas during the overall reaction (1-1-1), but is in equilibrium with the gas phase hydrogen and ammonia (see Chapter 1, Fig. 6),

$$N_2(a) + 3H_2(g) = 2NH_3(g) \qquad (3\text{-}1\text{-}13)\dagger$$

where (a) and (g) represent the adsorbed and gaseous states, respectively. If we designate the pressure of nitrogen which would be in equilibrium with the ambient hydrogen and ammonia by P_N^*,

$$P_N^* = P_A^2/(KP_H^3) \qquad (3\text{-}1\text{-}14)$$

the adsorbed nitrogen is in equilibrium with P_N^* during reaction (1-1-1).

The dependence of θ_N upon nitrogen pressure under the conditions of adsorption equilibrium is obtained by equating r_a and r_d in equations (3-1-12):

$$\begin{aligned} k_a P_N^e \exp(-g\theta_N^e) &= k_d \exp(h\theta_N^e) \\ \theta_N^e &= \frac{1}{g+h} \ln\left(\frac{k_a}{k_d} p_N^e\right) \end{aligned} \qquad (3\text{-}1\text{-}15)$$

where θ_N^e and P_N^e represent θ_N and P_N in adsorption equilibrium. As θ_N during the reaction, θ_N^r, is in adsorption equilibrium with P_N^*, from equations (3-1-14) and (3-1-15),

$$\begin{aligned} \theta_N^r &= \frac{1}{g+h} \ln\left(\frac{k_a}{k_d} P_N^*\right) \\ &= \frac{1}{g+h} \ln\left(\frac{k_a}{k_d} \frac{P_A^2}{KP_H^3}\right) \end{aligned} \qquad (3\text{-}1\text{-}16)$$

Under these circumstances, as the rates of nitrogen adsorption and desorption

† $N_2(a)$ represents the adsorbed species of nitrogen and does not imply that the adsorbed state is necessarily in molecular form.

(over the surface covered by nitrogen to the extent θ^r_N) are rate-determining, the rate of the ammonia synthesis, V_+, may be obtained as follows:

$$V_+ = k_a P_N \exp(-g\theta^r_N)$$

$$= k_a P_N \exp\left[-\frac{g}{g+h}\ln\left(\frac{k_a}{k_d}\frac{P_A^2}{KP_H^3}\right)\right]$$

$$= k_a\left(\frac{k_a}{k_d}\right)^{-g/(g+h)} K^{g/(g+h)} P_N (P_H^3/P_A^2)^{g/(g+h)} \tag{3-1-17}$$

$$= kP_N(P_H^{3\alpha}/P_A^{2\alpha})$$

where $\alpha = g/(g+h)$

In a similar manner, the rate of ammonia decomposition is

$$V_- = k_d \exp(h\theta^r_N)$$

$$= k_d(k_a/k_d)^{h/(g+h)} K^{-h/(g+h)} (P_A^2/P_H^3)^{h/(g+h)} \tag{3-1-18}$$

$$= k'(P_A^{2(1-\alpha)}/P_H^{3(1-\alpha)})$$

This rate equation (3-1-18) is in complete agreement with those given by equations (1-2-5) and (1-2-6) if $1 - \alpha$ is taken as 0·3 and 0·5, respectively. Such a treatment to explain the mechanism of ammonia synthesis and decomposition is called the Temkin–Pyzhev mechanism.[6]

The values of P_N^*, the pressure of nitrogen which would be in equilibrium with the adsorbed nitrogen, may be estimated under the reaction conditions as follows. Suppose ammonia decomposition is taking place when the ambient pressures of ammonia and hydrogen are 100 and 1 Torr, respectively. The equilibrium constant K'

$$K' = P_A/(P_N^{\frac{1}{2}}P_H^{\frac{3}{2}})$$

is $1{\cdot}27 \times 10^{-2}\,(\text{atm}^{-1})$ at 400°. Accordingly, P_N^* can be estimated from equation (3-1-14) to be of the order of 10^{11} atm at 400°C. If P_A and P_H are 1×10^{-5} and 1×10^{-7} Torr, respectively, P_N^* becomes $4{\cdot}5 \times 10^{17}$ atm. At higher temperatures K becomes smaller and P_N^* becomes higher correspondingly. The ammonia decomposition is being carried out at normal or low pressures, but the pressure of nitrogen that would be in equilibrium with θ^r_N

turns out to be as high as 10^{11} and 10^{17} atm. This is due to the *dynamic* properties of the overall reaction, where the chemical potential of the reaction intermediate preceding the rate-determining step is equal to that of the reactants, and it is the free energy drop (or the driving force) of the overall reaction which causes the adsorbed nitrogen to have such a high P_N^* value.† This difference between P_N^r and P_N^* is one of the characteristic properties of the dynamic behaviour of the chemical reaction, and cannot be observed from the static behaviour. As is easily understood in this case, the chemical potential of the adsorbed nitrogen is markedly different from that of the ambient nitrogen. This may be demonstrated by nitrogen adsorption measurements during the course of the reaction, or by using θ_N^r to estimate the chemical potential or P_N^* value of nitrogen over the iron surface in its working state. This is one of the reasons that the importance of adsorption measurements during surface catalysis has been emphasized for the elucidation of heterogeneous catalysis mechanisms.[7]

In heterogeneous catalysis, the amounts of each chemisorbed species on the catalyst surface may be measured, both during the catalytic reaction and under equilibrium conditions by means of various techniques which will be described later. This information can be used to determine the chemical potentials of each chemisorbed species under the working conditions of the catalyst.

Another example of such a kinetic treatment is given in an exercise on the decomposition of ammonia in a physical chemistry textbook. The question is, "The decomposition of ammonia on tungsten is a zero-order reaction. What does it mean?", and the answer given is as follows: "It shows that the catalyst surface is fully covered by ammonia molecules such that the amount of ammonia adsorbed on the surface remains constant being independent of ammonia pressure." Of course, the answer is one of several possibilities, but as will be discussed later in more detail, it is not the correct answer.

In a zero order reaction the catalyst sites on the surface may be saturated with an adsorbed species, coverage being independent of the ambient pressure, but this gives no information as to what species actually covers the sites. One of the direct approaches to elucidate the true reaction mechanism is to study directly the adsorbed species which stays on the catalyst sites during the course of the reaction. By measuring the amount of adsorbed species which saturates the catalyst sites, the number of active sites or the fraction of active region on the catalyst surface may be estimated. The first example where adsorption has been studied during the course of a catalytic reaction is the decomposition of germane on a germanium surface, a zero order reaction.

† It is not easy to prepare iron nitride on an iron surface by using nitrogen gas, but when ammonia decomposes under low hydrogen pressures, iron nitride may be formed under the reaction conditions because of the high P_N^* value.

3-2 The catalytic decomposition of germanium hydride

The reaction kinetics. In the field of heterogeneous catalysis it is generally accepted that, as the surface properties are usually sensitive to impurities, the cleanliness of the catalyst surface is of great importance in obtaining a well-defined reproducible catalytic system. The decomposition of hydrides such as arsine, stibine and germane over surfaces of their constituent elements, however, proceeds on clean surfaces which are constantly renewed by fresh deposition of the catalyst.

$$GeH_4 \rightarrow Ge + 2H_2 \tag{3-2-1}$$

The mechanism of such a well defined catalytic reaction has been studied in considerable detail.[8,9]

Reaction (3-2-1) proceeds over a Ge surface at temperatures above 200°C and the reaction is zero order with respect to both germane and hydrogen, being independent of their pressures. The partial pressure of GeH_4 falls linearly with time until the completion of the reaction (Fig. 2). At constant

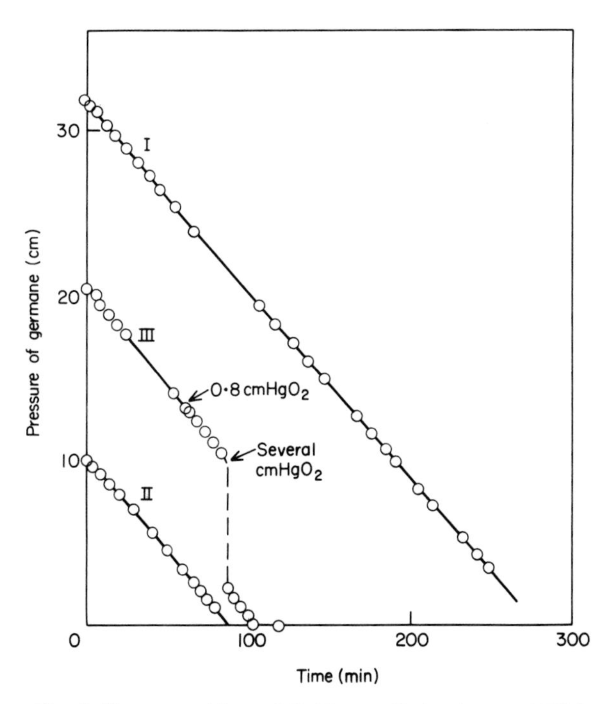

Fig. 2. Decomposition of GeH_4 on Ge surface at 278°C.

temperature, the gradient of the straight line is independent of the initial partial pressure of germane. The apparent activation energy is 41·2 kcal/mol and the rate of GeD_4 decomposition is slower than that of GeH_4, the ratio of the two decomposition rates being 1: 1·8 at 218°C.

The mechanism of reaction (3-2-1) has also been studied by means of isotopic techniques. When a mixture of GeH_4 and GeD_4 decomposes on a Ge surface, (i) little, if any, rearrangement occurs to form partially deuterated hydrides during the reaction,

$$\text{(i)} \qquad GeH_4 + GeD_4 \nrightarrow GeH_xD_y \qquad (x + y = 4) \qquad \text{(3-2-2)}$$

and (ii) there is considerable production of hydrogen deuteride during the decomposition to give an almost equilibrium distribution of H_2, HD and D_2:

$$\text{(ii)} \qquad GeH_4 + GeD_4 \rightarrow Ge + H_2 + 2HD + D_2 \qquad \text{(3-2-3)}$$

However, (iii) when germanium hydride decomposes in the presence of deuterium, no hydrogen deuteride is formed during the reaction:

$$\text{(iii)} \qquad GeH_4 + D_2 \rightarrow Ge + 2H_2 + D_2 \qquad \text{(3-2-4)}$$

but (iv) after the decomposition is over, the H_2–D_2 exchange reaction takes place to form HD:

$$\text{(iv)} \qquad H_2 + D_2 = 2HD \qquad \text{(3-2-5)}$$

Since an equilibrated mixture of H_2, D_2 and HD is produced by the simultaneous decomposition of GeH_4 and GeD_4, and since, as illustrated by (3-2-4), no H_2–D_2 exchange takes place during the decomposition reaction, the "hydrogen" must be produced intermolecularly. That is, reactions such as

$$GeH_4 \rightarrow GeH_2(a) + H_2$$

or

$$GeH_n \rightarrow GeH_{n-2}(a) + H_2 \qquad \text{(3-2-6)}$$

do not occur.

The adsorption of hydrogen on germanium. The chemisorption of hydrogen on germanium films, prepared by the decomposition of germane, has been studied. The adsorption is activated and reversible and the amounts of adsorbed hydrogen on the germanium surface at various temperatures are shown in Fig. 3. The activation energy for the initial adsorption was

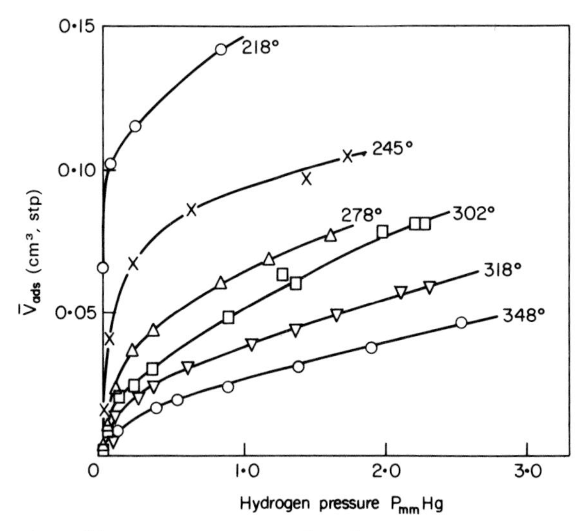

Fig. 3. Adsorption of hydrogen on germanium (temperatures shown are in °C).[8]

14·6 kcal/mol.† The adsorption may be expressed by Freundlich isotherms, which, when extrapolated on a log–log plot, converge to a common point (0·40 cm^3(stp)). At lower pressures, the adsorption may be approximately represented by the Langmuir isotherm for dissociative adsorption into two fragments (3-2-7), as shown in Fig. 4.‡

$$\frac{\theta}{1-\theta} = \sqrt{\mathbf{a}p} \qquad (3\text{-}2\text{-}7)$$

The heat of adsorption is 23·5 kcal/mol at $\theta = 0{\cdot}05$ and 23·3 kcal/mol at $\theta = 0{\cdot}10$. Saturated adsorption on the film was $1{\cdot}1 \times 10^{19}$ H_2 molecules (0·40 cm^3(stp)) on about $2{\cdot}15 \times 10^{19}$ surface germanium atoms, estimated from BET results and knowledge of the crystal structure of germanium. That is, saturated adsorption corresponds to one hydrogen atom per surface germanium atom.

Statistical mechanism gives an approximate value of $\sqrt{\mathbf{a}}$ in the Langmuir isotherm (3-2-7) from the partition functions of the hydrogen molecule. The

† The activation energy for the initial adsorption on an evaporated germanium film was subsequently measured by Bennett and Tompkins and found to be 16·6 kcal/mol at 273–296K.[10]

‡ Dissociative adsorption is also supported by the occurrence of the exchange reaction between H_2 and D_2 to form HD at the adsorption temperatures.

observed value of **a** at 318°C from Fig. 4, for example, is 0·11 $mmHg^{-\frac{1}{2}}$, while using a heat of adsorption of 23·5 kcal/mol the calculated value is 0·12 $mmHg^{-\frac{1}{2}}$. The two values are in satisfactory agreement, which suggests that the adsorption is immobile and dissociative as had been assumed in the calculation.

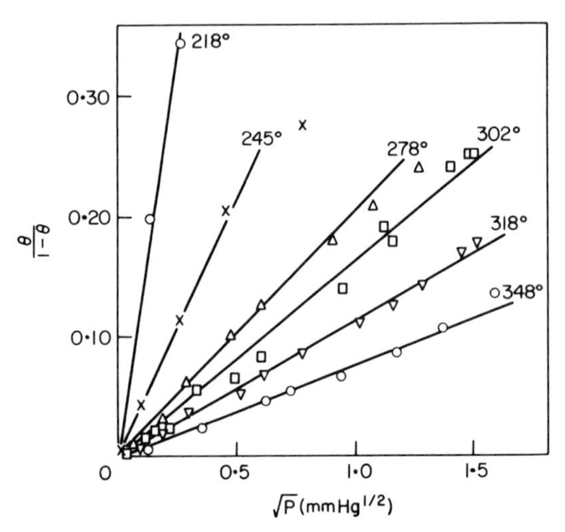

Fig. 4. Langmuir isotherms of hydrogen adsorption on a germanium surface.[8]

The fact that saturation corresponds to one adsorbed hydrogen atom per surface germanium atom demonstrates that all the germanium atoms on the surface take part in the chemisorption. As the initial adsorption behaviour is what would be expected from immobile dissociative adsorption on a uniform surface, all the germanium atoms on the surface may be considered to be homogeneous in their behaviour for the hydrogen adsorption. The uniform behaviour of germanium surfaces prepared by the decomposition of germane has also been demonstrated in ammonia adsorption.[13]

Dynamic treatments of the GeH_4 *decomposition.* The zero order kinetics of GeH_4 decomposition on a germanium surface strongly suggests a decomposition process on a surface fully covered with GeH_x radicals, the rate-determining step being the desorption or decomposition of these radicals. It is accordingly of interest to identify these radicals by studying more directly the behaviour of hydrogen on the germanium surface.

Germane was introduced into the reaction vessel at 250°C, after the decomposition reaction had proceeded for some time, the vessel was rapidly cooled down to liquid nitrogen temperature and the hydrogen in the gas phase pumped out. The reaction vessel was then brought to the temperature of solid carbon dioxide, which is higher than the boiling point of germane, and all the germane in the gas phase was evacuated from the vessel. The vessel was then warmed up to room temperature and after one night the pressure in the vessel was still less than 10^{-3} mmHg, which showed that practically all of the hydrogen and germane was removed from the reaction system, except those chemisorbed on the germanium surface. The reaction vessel was put in a naphthalene vapour bath and the pressure rise with time due to the desorption of chemisorbed hydrogen was followed with a McLeod gauge as shown by curve I in Fig. 5. No germane was desorbed on this occasion. The hydrogen desorption was initially rapid, the rate gradually decreasing to give an apparently constant amount of desorption at a fixed temperature. When the temperature was raised to 278°C, curve II in the figure was obtained, and when the temperature was lowered to 218°C again, curve III was obtained.

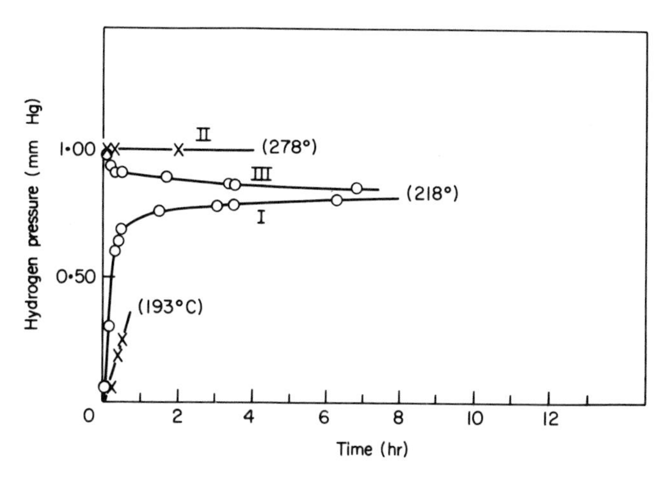

Fig. 5. Desorption and adsorption of hydrogen on germanium surface.[8]

The McLeod gauge was disconnected from the reacting system and pumped out, then hydrogen from the reacting system was reintroduced into the pressure gauge. By repeating this procedure, the equilibrium pressure could be extrapolated to zero as demonstrated in Fig. 6, and the total amount of hydrogen adsorbed on the germanium under the reaction conditions could be estimated. The amount of germane that had been allowed to decompose before the reaction vessel was cooled down was varied between 1·0 to 24·0 mm Hg and the final germane pressure was between 15 and 40 cm Hg. All the experiments showed that approximately the same amount of hydrogen was chemisorbed (e.g. 0·40, 0·39, 0·40 and 0·38 cm^3(stp)), and this is the same as the amount at adsorption saturation as obtained from the adsorption isotherm. If the amount of adsorbed hydrogen was in equilibrium with the ambient hydrogen pressure (1–24 mmHg), it is estimated that at a maximum, θ would be approximately 0·6.

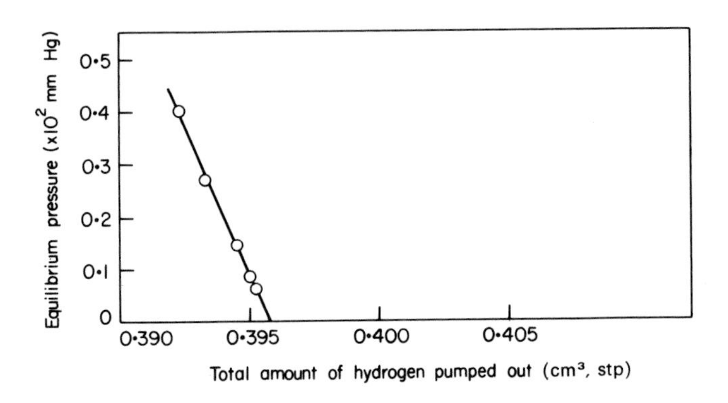

Fig. 6. Total amount of hydrogen taken out and corresponding equilibrium pressure at 278°C.[8]

Next, the rate of hydrogen desorption from a saturated surface was compared with the rate of hydrogen production during the germane decomposition reaction. The initial rate of hydrogen desorption from a saturated germanium surface was measured at 193°C, and as shown in Fig. 5, was $3{\cdot}0 \times 10^{-3}$ cm^3(stp)/min. On the other hand, the rate of germane decomposition on the surface was $1{\cdot}4 \times 10^{-2}$ cm^3(stp)/min at 218°, which corresponds to a rate of hydrogen production of $2{\cdot}8 \times 10^{-2}$ cm^3(stp)/min. From the activation energy of 41·2 kcal/mol this rate corresponds to $2{\cdot}9 \times 10^{-3}$ cm^3(stp)/min at 193°, agreeing very well with the rate of hydrogen desorption from Fig. 5.

The results obtained on the decomposition of germane on a germanium surface can be summarized as follows:

(1) The decomposition of germane is a zero order reaction, being independent of the pressure of germane and hydrogen and having an activation energy of 41·2 kcal/mol.

(2) During the course of the reaction the entire surface of the germanium is virtually covered by adsorbed hydrogen, one adsorbed hydrogen atom corresponding to one surface germanium atom, even though the equilibrium

hydrogen coverage at the ambient hydrogen pressures studied would be of the order of $\theta = 0{\cdot}4\text{–}0{\cdot}6$ at the most.

(3) The rate of hydrogen desorption from a surface saturated with adsorbed hydrogen is the same as the rate of hydrogen production during germane decomposition.

(4) The decomposition of germane in the presence of D_2 produces no HD during the course of the reaction, whereas after all the germane has decomposed, H_2 and D_2 start to react to form HD.

(5) The decomposition of a mixture of GeH_4 and GeD_4 produces H_2, HD and D_2 in almost equilibrium proportions, while no GeH_xD_y formation takes place.

(6) The ratio of the rates of decomposition of germane and deuterogermane is 1·8: 1.

(7) The adsorption of hydrogen on a germanium surface is reversible with an activation energy of 14·6 kcal/mol for the initial adsorption. It obeys a Langmuir isotherm for dissociative immobile adsorption at lower coverages, $\theta = \sqrt{\mathbf{a}p}$, and saturated adsorption corresponds to one hydrogen atom per surface germanium atom, suggesting that all the surface germanium atoms participate in the adsorption.

The zero-order kinetics implies that the active surface sites are saturated with GeH_x during the course of the reaction and adsorption measurements during the reaction demonstrated that the species is GeH, as there is a one to one correspondence between the number of adsorbed hydrogen atoms and surface germanium atoms.

The adsorption measurements during the reaction which showed that the surface was saturated with hydrogen, demonstrated that the chemical potential of the hydrogen adsorbed on the surface during the reaction, is markedly higher than that of the ambient hydrogen gas. This potential difference is also supported by observation (4) which revealed that the gas phase hydrogen cannot react with the surface during the decomposition reaction as the surface is already saturated with hydrogen from the decomposition of germane.

It is striking that the rate of hydrogen desorption from a saturated surface is equal to the rate of the decomposition reaction at the same temperature as was observed in (3). Accordingly it can be concluded that the catalyst surface is saturated with adsorbed hydrogen and that desorption from the saturated surface is the rate-determining step.

The slower rate (by a factor of 1·8) of the decomposition of deutero-germane shows that deuterium desorbs more slowly than hydrogen from a germanium surface. This is consistent with the normal zero-point energy difference for the hydrogen isotopes, which is to be expected from a bond breaking.

The lack of HD formation during the decomposition of GeH_4 in the presence of D_2 (observation 4) rules out an Eley–Rideal mechanism, involving collisions of gas phase or physically adsorbed deuterium molecules with the chemisorbed species on the surface.

The decomposition of germane, accordingly, takes place by splitting off hydrogens one by one:

$$GeH_4(g) \rightarrow GeH_3(a) + H(a)$$

$$GeH_3(a) \rightarrow GeH_2(a) + H(a)$$

$$GeH_2(a) \rightarrow GeH(a) + H(a)$$

$$GeH(a) \rightarrow Ge + \tfrac{1}{2}H_2(g) \quad (\text{or } H(a))$$

and the last step is the main rate-determining step of the reaction. However, if the desorption of hydrogen was the only rate-determining step, the chemisorption of the germane would become so strongly hindered by the chemisorbed hydrogen that the rate of germane chemisorption could become even slower than the rate of the last step. Consequently, with chemisorbed hydrogen covering most of the surface, another free energy drop may be involved at the chemisorption of germane step. The existence of this second potential energy drop is supported by observations (5), the lack of GeH_xD_y formation from a mixture of GeH_4 and GeD_4 during the decomposition reaction.

The activation energy for hydrogen desorption appears to be virtually independent of coverage, being 41·2 kcal/mol at saturation, and 38·1 kcal/mol from a bare surface (the initial heat of adsorption plus the activation energy for initial adsorption; 23·5 + 14·6). Accordingly the rate of the H_2–D_2 exchange reaction (3-2-5) may be estimated from the rate of desorption at full coverage, or half the rate of the decomposition of germane, r_d. If the hydrogen coverage in adsorption equilibrium is θ_H in Fig. 3, the rate of HD formation, r_e, from an equimolar mixture of H_2 and D_2 would be

$$r_e = r_d \theta_H^2 / 2$$

The calculation gives results in good agreement with experiments.

The catalytic decomposition of germane is a very suitable reaction to study. Only two elements are involved and, in addition, because the chemisorption of hydrogen on germanium is an activated process, the amount of chemisorption under the reaction conditions can be measured by rapidly cooling the catalyst. It is an example of how many different pieces of information can be put together to elucidate the reaction mechanism.[11–15]

3-3 The decomposition of ammonia on tungsten surfaces

The decomposition of ammonia on tungsten has been widely studied by many investigators as a typical catalytic reaction. Hinshelwood and Burk[16] reported that the reaction is zero order, the rate being independent of the ammonia, hydrogen and nitrogen pressures. However, as shown in Fig. 7, this is not the case in a strict sense. Hailes[17] later reported that the apparent activation for the decomposition depends upon the ammonia pressure, being greater at higher pressures.

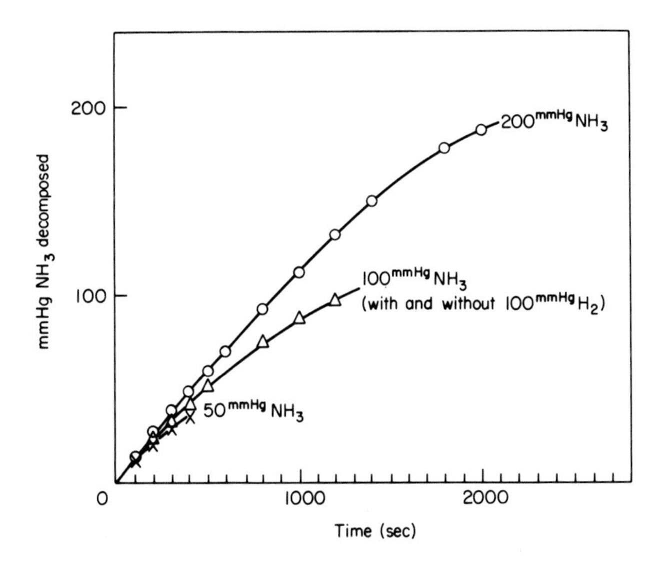

Fig. 7. Pressure–time curves for the decomposition of ammonia on tungsten at three different initial pressures.[16]

The zero order kinetics may be interpreted as indicating that the catalyst sites are saturated with ammonia molecules, and an example of a textbook in which this interpretation has been made was mentioned in 3-1.

However, Frankenburger and Holder[18] found that when ammonia was brought into contact with a tungsten surface, hydrogen formation took place even at 150°C. At 250°C, more than one molecule of hydrogen (or two hydrogen atoms) desorbed per adsorbed ammonia molecule, but no nitrogen molecules were formed.

$$NH_3 \rightarrow N(a) + \tfrac{3}{2}H_2(g)$$

Accordingly, at the usual temperatures of more than 600°C, it is improbable that the surface is saturated by adsorbed molecular ammonia. It is more plausible to assume that the catalyst surface is fully covered by chemisorbed nitrogen, its desorption being the rate-determining step. In this case, the rate of nitrogen desorption from the saturated surface would be expected to remain constant, being independent of the ambient partial pressures.

It is of interest to note that Jungers and Taylor,[19] and Barrer[20] studied the kinetic isotope effect in ammonia decomposition by employing NH_3 and ND_3 and observed that on a tungsten filament at around 1000 K NH_3 is decomposed 1·6 times as fast as ND_3. If the rate-determining step is the desorption of nitrogen from a surface saturated with nitrogen, NH_3 and ND_3 should decompose at the same rate, as all the hydrogen should be split off to form hydrogen molecules prior to the nitrogen desorption process. Accordingly, the kinetic isotope effect must be explained by some other reaction mechanism, for instance by assuming that NH(a) saturates the surface, but the experiments of Frankenburger and Holder make this improbable.

Under those circumstances, Tamaru[21] proposed to measure directly the adsorption on the catalyst surface during the course of the reaction in order to identify the surface species which fully covers the surface independent of the ambient pressures.

To measure the adsorption during the course of the reaction he used a closed circulating system and a tungsten powder catalyst with a large surface area which was carefully reduced by hydrogen. The results obtained by his approach were as follows:

Below 723 K, ammonia decomposed to produce only hydrogen, virtually all of the nitrogen remaining on the tungsten, with only a negligible amount of nitrogen molecules being formed. The initial rate of hydrogen production was independent of the ammonia pressure, but became pressure-dependent at later times.

At temperatures above 773 K, there was considerable nitrogen formation and, in contrast to lower temperatures, a negligible amount of hydrogen was adsorbed on the catalyst surface in any form. On the other hand, multilayer nitrogen adsorption occurred indicating the formation of surface nitride layers or the chemical potential of the chemisorbed nitrogen to be higher than that of nitrogen gas. The thickness of the surface nitride layer was determined from the balance between the rate of supply of chemisorbed nitrogen from the decomposing ammonia and its rate of desorption.

The rate of surface nitride formation decreased rapidly as the thickness of the layer increased, and this was more pronounced at lower temperatures, the apparent activation energy of nitride formation increasing as the nitride layers become thicker. The rate of desorption of nitrogen, i.e. the rate of decomposition of the surface nitride, on the other hand, increased rapidly as the nitride layer became thicker. Consequently, as the thickness of the nitride layer increased, the rate of supply of adsorbed nitrogen, which was apparently first order with respect to ammonia pressure, became slower, while the rate of nitrogen desorption became faster until the supply and the consumption of chemisorbed nitrogen were dynamically balanced.

The rate of nitrogen desorption measured separately in the absence of gas phase ammonia was compared with the rate of nitrogen production in the overall reaction at equal amounts of nitrogen sorption, and it was found that both rates were in reasonable agreement. It was thus demonstrated that nitrogen desorption is one, at least, of the rate-determining steps of the overall reaction.†

The kinetic isotope effect in the decomposition rate is also explained by this dynamic equilibrium mechanism. The thickness, or the chemical potential, of the surface nitrogen layer depends on the balance between the supply of nitrogen from the ammonia and its desorption. The different hydrogen isotopes in the ammonia would result in a difference in the chemical potential (thickness) of the surface nitrogen. This will influence the rate of nitrogen desorption in the overall reaction even though no hydrogen is being chemisorbed.

As discussed in Chapter 1, if nitrogen desorption is the rate-determining step, the equilibrium nitrogen pressure corresponding to the actual amount of chemisorbed nitrogen may become incredibly high, which is one of the reasons why the surface nitride is formed. For the decomposition of ammonia on tungsten, the amount of chemisorbed nitrogen results from a balance between the rate of supply and the rate of consumption. Therefore, the overall decrease in the chemical potential will be associated with both processes, and consequently the hypothetical equilibrium pressure of the chemisorbed nitrogen, although still very high, will be correspondingly less than if the nitrogen desorption was the only rate-determining step. This dynamic equilibrium mechanism for the ammonia decomposition explains the available data such as the zero order (or low fractional order) kinetics, high chemical potential of chemisorbed nitrogen and the kinetic isotope effect.

Matsushita and Hansen[22] studied the decomposition of ammonia on polycrystalline tungsten by flash-filament desorption spectroscopy. Ammonia was partially decomposed at 300 K resulting in the desorption of approximately 1·5 atoms of hydrogen per adsorbed ammonia molecule. The hydrogen desorption spectrum from the ammonia decomposition was qualitatively similar to that which would have resulted from a comparable amount

† These experimental results indicate that molecular nitrogen is formed by the desorption of nitride layers and not by a process in which the ambient gas participates such as $NH_3(g) + N(a)$.

of hydrogen dosed onto a clean surface. The desorption was substantially complete by 900 K while the nitrogen desorption started at 870 K with a peak around 1100 K. This desorption peak was called x-nitrogen and is almost completely resolved from the β-nitrogen peak as shown in Fig. 8.

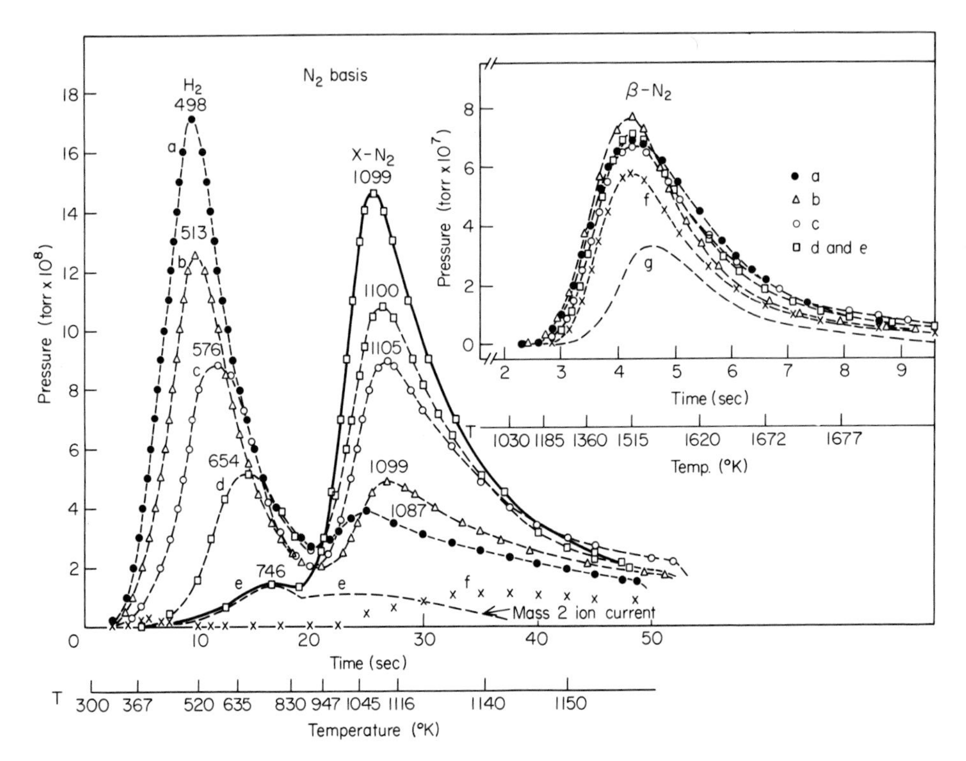

Fig. 8. Flash-desorption spectra showing the separation of hydrogen, x-, and β-nitrogen upon decomposition of ammonia.[22]

When the surface was predosed with $^{15}NH_3$ and subsequently dosed with $^{14}NH_3$, extensive mixing of the isomers was found in both the x- and β-nitrogen peaks, but with some bias in the x-nitrogen peak toward the predosed nitrogen isotope. They also suggested that the average rate of x-nitrogen desorption agrees well with literature values for the rate of ammonia decomposition on tungsten wires.

Although their dosing conditions were too limited to study the behaviour of nitrogen in the surface layers under the reaction conditions, the results they obtained are in reasonable agreement with those obtained by Tamaru over tungsten powder, giving a high chemical potential of chemisorbed nitrogen.

Dawson and Hansen,[23] and Peng and Dawson,[24] also studied ammonia decomposition and postulated surface species such as $W_2N(\beta)$, $WN(\delta)$ and $W_2N_3H(\eta)$.

McAllister and Hansen[25] studied the catalytic decomposition of ammonia over the (100), (110) and (111) crystal faces of tungsten. In all cases the rate of reaction could be expressed by the following equation,

$$R = A + BP_{NH_3}^{\frac{2}{3}} \qquad (3\text{-}3\text{-}1)$$

being independent of the partial pressures of nitrogen and hydrogen, as shown in Figs 9 and 10. The constants A and B varied with the crystal face, the value of B being largest for the (111) face. In the decomposition of NH_3 and ND_3 on the (111) face at 860 K, $A_{NH_3} = A_{ND_3}$ and $B_{NH_3} \approx 1{\cdot}47 B_{ND_3}$. They tried to explain their kinetic equation by assuming that the A term is due to the decomposition of WN which fully covers the surface, while the second term of equation (3-3-1) arises from interactions between the low coverage species $W_2N_3H_2$, WHN and W_2N. Of course such speculation is a game which may be played in many different ways, as has been discussed in 3-1.

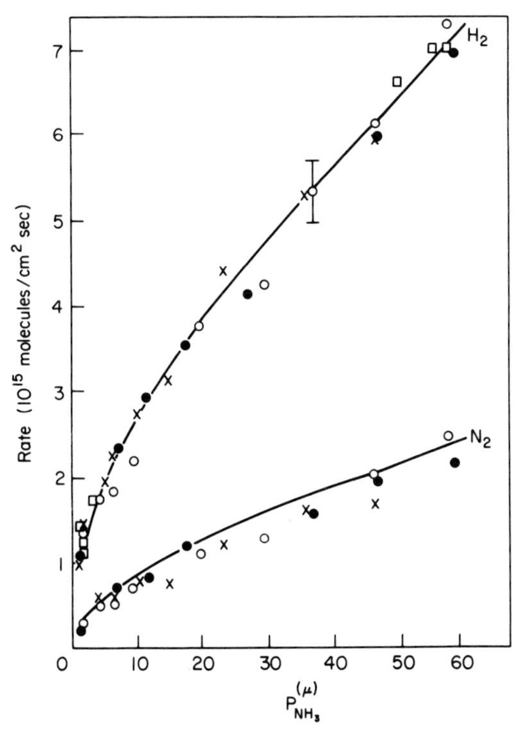

Fig. 9. Decomposition of ammonia on W(111) surface at 848 K.[25]

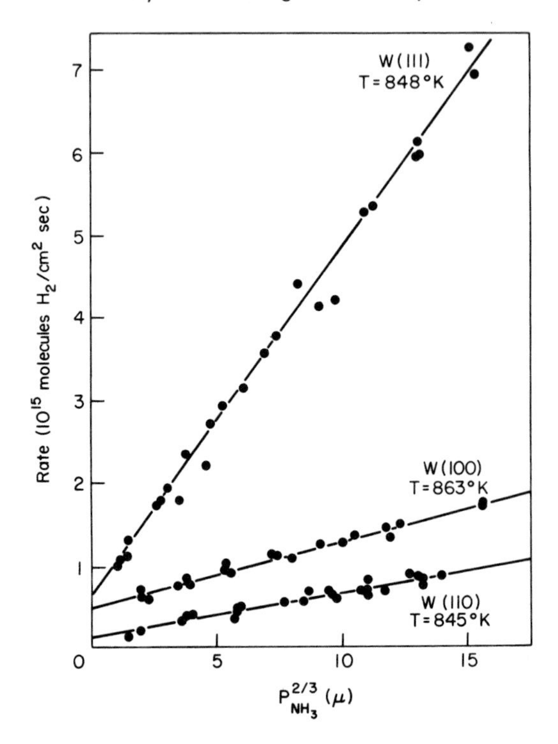

Fig. 10. Comparison of ammonia decomposition rates on W(111), W(100), W(110) faces.[25]

Although it is not clear how far we may discuss the reaction mechanism on the basis of the kinetic equation (3-3-1), it demonstrates that the reaction is zero order at very low ammonia pressures, and of a low (fractional) order at higher pressures. The former can be interpreted, as have McAllister and Hansen, to be due to nitrogen desorption from a saturated nitrogen monolayer, since the hydrogen exhibits no kinetic isotope effects. The saturated adsorption of nitrogen at low ammonia pressure suggests a chemical potential of chemisorbed nitrogen definitely higher than that of the ambient nitrogen gas, with nitrogen desorption being a rate-determining step. At higher ammonia pressures the chemical potential of the chemisorbed nitrogen will become higher than that of the saturated adsorption. The second term of equation (3-3-1), accordingly, may be interpreted as arising from nitrogen desorption from a multilayer surface nitride, as has been observed on tungsten powder. The chemisorbed nitrogen, in this case, operates as a buffer, being balanced between its supply and consumption, and causing the low (fractional) order of the reaction with respect to ammonia pressure that is observed.

If the reaction proceeds in such a manner, the behaviour of the nitrogen on the surface which is directly associated with the overall reaction is not the same as that of nitrogen chemisorbed from molecular nitrogen at ordinary pressures. Consequently, if the chemisorption of hydrogen and nitrogen under ordinary pressures and the chemisorption of ammonia at room temperature are studied, the experimental results will be almost irrelevant to the mechanism of ammonia decomposition over the catalyst. Further, the properties of the surface are not those of the metal surface in its ordinary state. Accordingly, we can understand the importance of adsorption and reactivity measurements of the surface species during the course of the reaction.

References

1. R. M. Barrer, *Trans. Faraday Soc.* **32,** 490 (1936).
2. H. W. Melville and H. L. Roxburgh, *J. Chem. Soc.* **1933,** 586.
3. L. Rheaume and G. Parravano, *J. Phys. Chem.* **63,** 264 (1959).
4. V. Grassi, *Nuovo Cimento,* **11,** (6), 147 (1916).
5. R. N. Pease, *J. Am. Chem. Soc.* **45,** 1196 (1923).
6. M. I. Temkin and V. Pyzhev, *Acta Physiochim. USSR,* **12,** 327 (1940).
7. K. Tamaru, *Adv. in Catalysis,* **15,** 65 (1964).
8. K. Tamaru, M. Boudart and H. Taylor, *J. Phys. Chem.* **59,** 801 (1955); P. J. Fensham, K. Tamaru, M. Boudart and H. Taylor, *J. Phys. Chem.* **59,** 806 (1955); K. Tamaru and M. Boudart, *Adv. in Catalysis,* **IX,** 699 (1957).
9. K. Tamaru, *J. Phys. Chem.* **61,** 647 (1957).
10. M. J. Bennett and F. C. Tompkins, *Trans. Faraday Soc.* **58,** 816 (1962).
11. K. Tamaru, *J. Phys. Chem.* **59,** 777 (1955).
12. K. Tamaru, *J. Phys. Chem.* **59,** 1984 (1955).
13. K. Tamaru, *J. Phys. Chem.* **60,** 612 (1956).
14. K. Tamaru, *J. Phys. Chem.* **60,** 610 (1956).
15. K. Tamaru, *Bull. Chem. Soc. Japan,* **31,** 666 (1958).
16. C. N. Hinshelwood and R. E. Burk, *J. Chem. Soc.* **127,** 1105 (1925).
17. H. R. Hailes, *Trans. Faraday Soc.* **27,** 601 (1931).
18. W. Frankenburger and A. Holder, *Trans. Faraday Soc.* **28,** 229 (1932).
19. J. C. Jungers and H. S. Taylor, *J. Am. Chem. Soc.* **57,** 679 (1935).
20. R. M. Barrer, *Trans. Faraday Soc.* **32,** 490 (1936).
21. K. Tamaru, *Trans. Faraday Soc.* **57,** 1410 (1961).
22. K. Matsushita and R. S. Hansen, *J. Chem. Phys.* **52,** 4877 (1970).
23. P. T. Dawson and R. S. Hansen, *J. Chem. Phys.* **48,** 623 (1968).
24. Y. K. Peng and P. T. Dawson, *J. Chem. Phys.* **54,** 950 (1971).
25. J. McAllister and R. S. Hansen, *J. Chem. Phys.* **59,** 414 (1973).

Chapter 4

The Application of Spectroscopic Techniques to the Dynamic Treatment of Adsorbed Species under Reaction Conditions

4-1 Infrared absorption spectroscopy and the elucidation of some mechanisms of heterogeneous catalysis

As has been emphasized in the preceding chapters, kinetic studies of reactions are not sufficient by themselves to elucidate the mechanisms of heterogeneous catalysis. As illustrated in Chapter 1 a certain kinetic expression may be interpreted on the basis of different mechanisms even in the simple case of uniform catalyst sites. In practice, the catalyst surface is usually not uniform and few chemisorption systems follow thermodynamically ideal behaviour such as the Langmuir adsorption isotherm. The occurrence of non-ideal adsorption behaviour is due in part to the fact that the adsorption itself influences the nature of the surface (e.g. the work function) on which subsequent adsorption takes place; this is called "induced heterogeneity". As well as that, many surfaces are energetically heterogenous: for example, a small perfect crystal of platinum of a size typically found in gasoline reforming catalysts, may contain 85 metal atoms, with 78% of them exposed at the surfaces of the crystal. The coordination number of 63% of the surface atoms is less than the normal value of 9 for the (111) octahedral faces: it is 7 for the 36 atoms along the edges and only 4 for the 6 corner atoms. Accordingly, the adsorbed surface species have a broad spectrum of reactivity, deviations from ideality are large and hence the interpretation of the rate laws is far from clear-cut.

An effective approach to the problem of identifying the real reaction intermediate and of determining the pathway of the reaction is first to detect the surface species by some means and then to study their dynamic behaviour under the reaction conditions. Unfortunately, many instrumental techniques such as high-resolution NMR, are powerful tools for homogeneous systems, but are not readily adaptable for studying heterogeneous systems. Some techniques such as LEED, esr and Mössbauer spectroscopy are limited to a particular form or kind of catalytic system and some, e.g. electron spectroscopy, require low ambient gas pressures.

Infrared spectroscopy of adsorbed species, a technique pioneered by Terenin,[1] Eischens and their co-workers,[1] can detect and identify surface species in many catalytic systems under normal reaction conditions. It also enables the dynamic behaviour of the adsorbed species to be followed by means of isotopic tracers under the stationary-state of the overall reaction. This, as explained previously, is an effective approach to the problem of identifying the reaction pathway. For example, the behaviour of chemisorbed species under reaction conditions may be studied as these conditions are rapidly changed, for instance, by replacing one of the reactants or products by its isotopic analogue or by changing the reaction temperature or the pressure of one of the reaction species. Infrared spectroscopy may be employed to determine the kind and structure of the chemisorbed species on the catalyst surface during the reaction, and to measure their surface concentrations at the same time as the rate of the overall reaction is being measured. These observations may be carried out under various reaction conditions to examine the dependence of the reaction rate upon the amounts of each chemisorbed species as well as those of the reactants in the ambient gas. The nature of this dependence gives information as to which chemical species participate in the rate-determining step of the overall reaction.

Experimentally, the above approach may be carried out as follows. An apparatus suitable for dynamic measurements and illustrated in Fig. 1 consists of a sampling manifold (A) and a closed circulating system (B). The latter is composed of one or two infrared cells (g), an all-glass circulating pump (c), a cold trap (e), and a manometer (d). The whole apparatus is movable so that the infrared cell may be placed in position in the beam of an infrared spectrometer. If necessary, an identical cell can be placed in the reference beam of a double beam instrument to compensate for the background spectrum of the catalyst, emitted radiation from the hot sample, and the absorption spectra of the gas phase. The reaction cell is *similar* to those employed in the ordinary static method being equipped with a sample holder, a microheater, a thermocouple pocket, and infrared transparent windows (Chapter 2, Fig. 19). The catalyst samples may be supported metals, pressed pellets, evaporated films or other samples prepared in the same way as samples for the ordinary infrared methods.

A known amount of reactant gas is introduced into the reaction system and is circulated through the cell which is at the desired temperature. In the course of the reaction the behaviour of the adsorbed species is measured at suitable time intervals under various reaction conditions. In some cases a relatively large amount of catalyst under the same reaction conditions is included in the

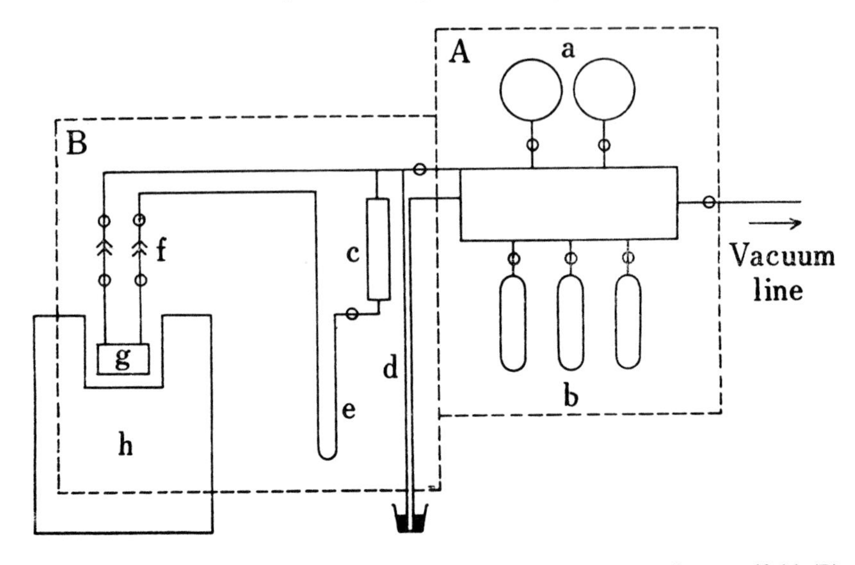

Fig. 1. Apparatus for the dynamic infrared measurements. (A), Sampling manifold: (B), closed circulating reaction system (a), Gas reservoirs: (b), sample bottles: (c), circulating pump: (d), mercury manometer: (e), trap: (f), ball joints: (g), infrared cell: (h), infrared spectrometer.[19]

circulating system so that the amount of adsorption can also be measured volumetrically during the reaction.

In the steady-state of the reaction the circulating reactant may be quickly replaced by an isotopically labelled molecule and the behaviour of the isotope in the adsorbed species and the reaction products is constantly watched. The behaviour of the adsorbed species under the reaction conditions can be followed by the isotope labelling, and be compared with the simultaneously measured overall reaction rate. Thus the real reaction intermediate may be identified. As described in Section 1-4, by examining the dependence of the reaction rate upon the amounts of adsorbed species and the partial pressures of reactants, we may also elucidate the reaction mechanism, e.g. Langmuir–Hinshelwood or Eley–Rideal.

If only a limited part of the catalyst surface is active for a particular reaction, the adsorption on the total catalyst surface may have little to do with the catalytic reaction. However, if the reaction takes place through a fraction of the adsorbed species, the isotope passes from reactant to product through part of the adsorbed species only, and this may be identified by examining the behaviour of the adsorbed species during the course of the reaction.

In this case, however, if the mobility of the surface species is high so that there is isotope mixing among all of the surface species, some ambiguity is inevitably involved. Although the isotope labelling technique is effective in many cases, exchange reactions which may take place on the side can cause the results to be ambiguous and an example of this will be seen later in the decomposition of formic acid on alumina.

4-2 The H_2 -D_2 exchange reaction over ZnO

The chemisorption of hydrogen in zinc oxide takes place readily at various temperatures as demonstrated in Fig. 2. The infrared spectra of the adsorbed hydrogen and deuterium on zinc oxide have bands at 3490 (ZnOH), 2584

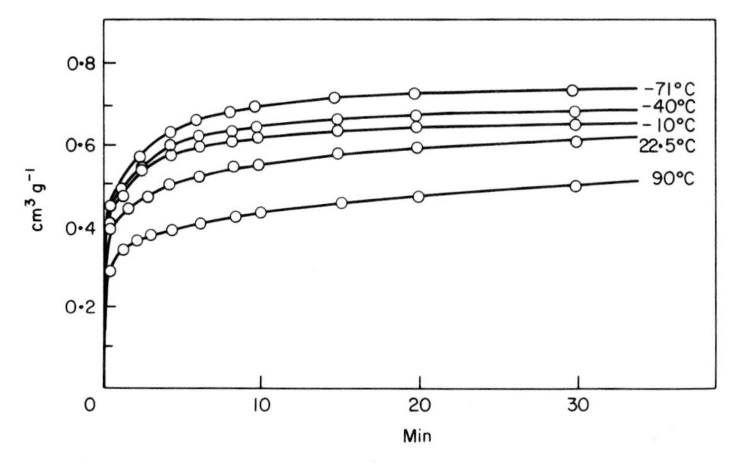

Fig. 2. The rate of hydrogen chemisorption over ZnO at various temperatures.[6]

(ZnOD), 1712 (ZnH) and 1233 (ZnD) cm^{-1} (Chapter 2, Fig. 21) as was reported by Eischens, Pliskin and Low in accordance with the equation:[2]

$$H_2 + Zn{-}O = H{-}Zn{-}O{-}H$$

At low temperatures, there is an additional component that cannot be assigned to dissociate hydrogen adsorption. Figure 3 shows the band observed at 2885 cm^{-1} when deuterium is adsorbed on zinc oxide at $-195°C$.[3] Table I gives the positions of the low temperature bands for H_2, D_2 and HD adsorption along with the calculated frequencies for the fundamental vibrations of the gaseous molecular species. The observed bands are obviously

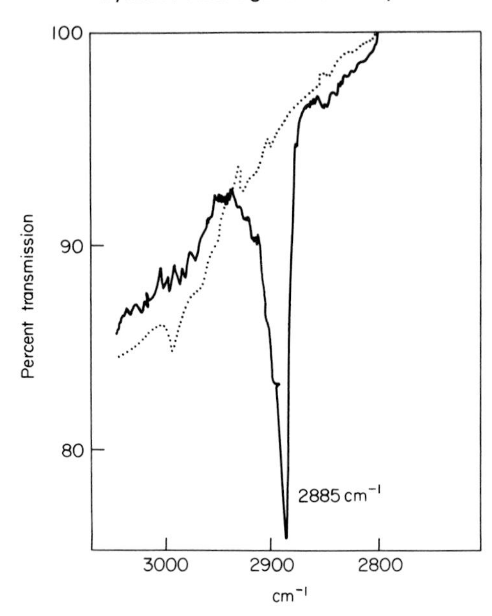

Fig. 3. Spectrum of molecular deuterium on zinc oxide at $-195°C$. The dotted line represents the background for the degassed catalyst.[3]

due to the molecular species with the frequencies shifted because of interactions with the surface. In the gas phase, infrared transitions due to the vibrations of homonuclear diatomic molecules are forbidden, but it has been observed that such transitions may be induced by strong electric fields.[4] Consequently these bands may be ascribed to transitions made possible by a high effective field at the surface.[5] The molecular hydrogen is only weakly adsorbed and may be desorbed by evacuation at $-195°C$ or by the adsorption

Table I. Vibration frequencies for molecular hydrogen

Species	ω^{22}(gas), cm^{-1}	ω(adsd), cm^{-1}	$\Delta\omega$, cm^{-1}	$(\Delta\omega/\omega) \times 100$
H_2	4161	4019	142	3·41
HD	3627	3507	120	3·31
D_2	2990	2887	103	3·44
				Av 3·39 ± 0·05

of water vapour. It has also been reçognized that the molecular adsorption takes place on the same sites as the dissociative adsorption, because if the zinc oxide is poisoned with water, the bands for both types of adsorption disappear.

The results of volumetric adsorption measurements during the H_2–D_2 exchange reaction are shown in Fig. 4.[6] Here, H_2 was preadsorbed onto the zinc oxide and subsequently D_2 was added to the system. As the exchange

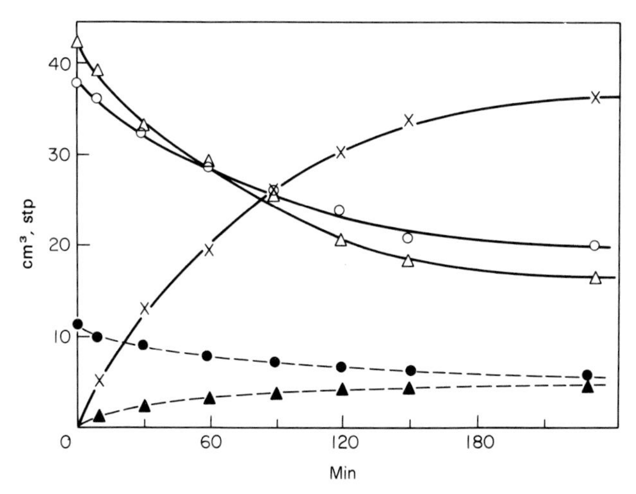

Fig. 4. Adsorption measurements during the H_2-D_2 exchange reaction over ZnO by the volumetric method. H(a) = 22 cm^3 (stp) preadsorbed (−49°C). (○) H_2(g); (●) $\frac{1}{2}$H(a); (△) D_2(g); (▲) $\frac{1}{2}$D(a); (×) HD(g).[6]

proceeded, preadsorbed H(a) was displaced by D(a), the rate of HD formation obeying a first order rate equation with no induction period. Figure 5 shows infrared spectroscopic results similar to the results in Fig. 4 except that in this case, the D_2 was preadsorbed. The rate of decrease in the intensities of the ZnD and ZnOD bands was equal to the rate of increase in the intensities of the ZnH and ZnOH bands, the total amounts of each species, ZnH+ ZnD, and ZnOH+ ZnOD, staying constant. Their exchange rate also obeyed the first order rate equation.

The dependence of the rate of exchange upon the pressure was also studied by the volumetric method as well as the infrared technique, which revealed that the rate is proportional to the partial pressure of H_2 (or D_2). The rate of the ZnH disappearance was proportional to the partial pressure of D_2. The rate of ZnOH disappearance was also proportional to the pressure of D_2 although the two species had different rate constants. These results strongly suggest that

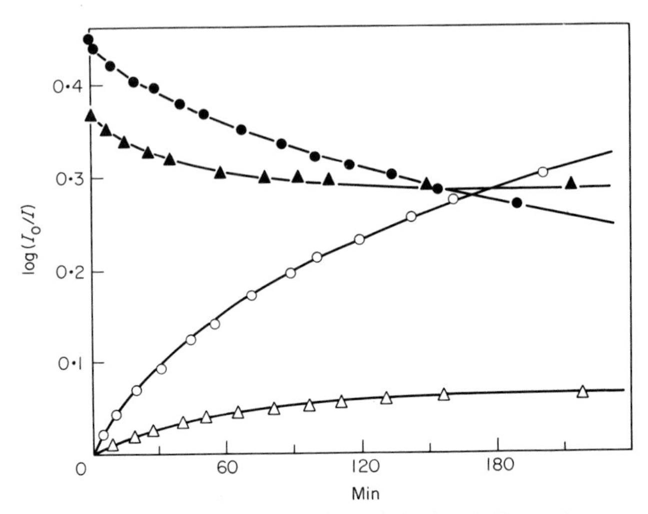

Fig. 5. Changes in the intensities of infrared bands in the H_2-D_2 exchange reaction over ZnO. D_2 was preadsorbed and then H_2 was introduced (0°C). (○) ZnH; (△) ZnOH; (●) ZnD; (▲) ZnOD.[6]

the exchange reaction takes place when the chemisorbed hydrogen observed in the infrared spectra reacts with molecular deuterium either in the ambient gas or weakly adsorbed on the catalyst surface (Eley–Rideal mechanism).

To confirm the above interpretation, the exchange reaction between the chemisorbed species and "hydrogen" molecules in the gas phase was examined

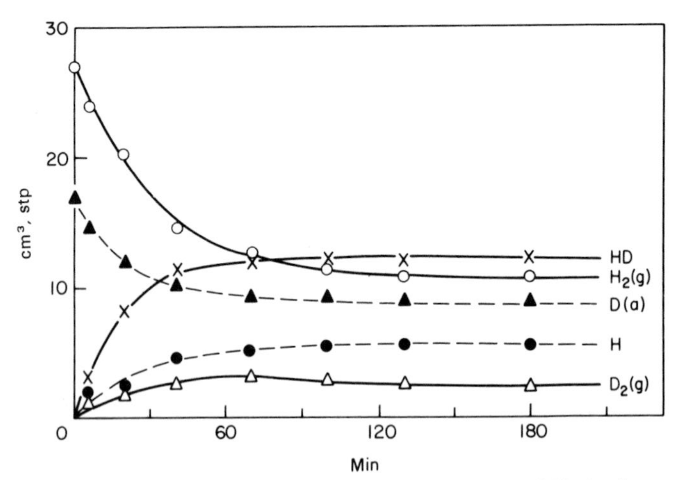

Fig. 6. The exchange reaction between chemisorbed D(a) and H_2 in the ambient gas (−60°C). (○) H_2(g); (●) H(a); (△) D_2(g); (▲) D(a); (×) HD(g).[6]

in more detail. D_2 (or H_2) was preadsorbed onto the zinc oxide surface at $-60°C$, the D_2 in the gas phase replaced by H_2, and the rate of HD formation was followed. If the exchange reaction proceeds with repetitive dissociative chemisorption and desorption involving the chemisorbed species, H(a) and D(a), the exchange rate should be proportional to the amounts of H(a) and D(a) and there must be an induction period for HD formation because at the beginning of the reaction only one of the hydrogen isotopes is present on the surface. The experimental results indicated that this is not the case, as demonstrated in Fig. 6. The rate of HD formation obeyed the first order rate equation from the beginning of the reaction.

4-3 The isomerization and hydrogenation of olefins over zinc oxide

The isomerization of olefins over metal oxides such as zinc oxide does not require the presence of hydrogen as a co-catalyst, indicating that hydrogen addition across the double bond does not play a role in the reaction. However, it is considered that the dissociation of hydrogen from the olefin plays an important role in reactions of this type. Burwell and his co-workers[7] suggested that allylic species may be reaction intermediates for isomerizations over chromia:

$$R-CH_2-CH{=}CH-R' \rightarrow R-\underset{*}{CH}-CH{=}CH-R' + \underset{*}{H} \rightarrow$$

$$R-CH{=}CH-\underset{*}{CH}-R' + \underset{*}{H} \rightarrow R-CH{=}CH-CH_2-R'$$

Such allylic species are also considered to be intermediates in the oxidation of olefins.[8]

Dent and Kokes chose propylene for their prototype olefin because it is the simplest olefin capable of forming an allylic species.[9]

When hydrogen is adsorbed onto zinc oxide (Kadox 25) in the presence of ethylene, the OH band shifts upwards in frequency by about 25 cm^{-1} and the ZnH band shifts downwards by 50 cm^{-1}. When propylene is preadsorbed on the catalyst and hydrogen is subsequently introduced, no additional bands near 3500 and 1700 cm^{-1} are observed, demonstrating that the preadsorbed propylene blocks off nearly all of the sites for ZnH and OH formation. Ethylene is adsorbed to form a π-complex which is readily removed by brief pumping at room temperature.[10] Propylene also adsorbs to form a π-complex with similar characteristics, but in addition forms another species which is removable only by evacuation at 125°C for about 1 hour. This demonstrates that ethylene and propylene adsorption on zinc oxide are drastically different and that chemisorbed propylene is much more strongly bonded to the surface than is ethylene.

The absorption spectra of the chemisorbed propylene showed C—H and O—H stretching vibrations. No band assignable to ZnH was observed, thus suggesting that propylene is adsorbed on ZnO by dissociation into a propyl group and a hydrogen atom as

$$\begin{array}{cc} C_3H_5 & H \\ | & | \\ -Zn & -O- \end{array}$$

The frequency of the OH band is about 105 cm^{-1} higher than that observed from adsorbed hydrogen.

Five bands, attributable to C—H stretching vibrations, are observed in the region near 3000 cm^{-1} and bands observed between 1450 and 1200 cm^{-1} can be assigned to C—H deformation vibrations. A band observed at 1545 cm^{-1} might be interpreted as the 1652 cm^{-1} propylene double-bond stretching shifted by 107 cm^{-1}, suggesting that π-bonding is involved in the chemisorption.

When butene-1 was introduced onto ZnO, a strong surface hydroxyl band at 3615 cm^{-1} was shifted about 5 cm^{-1} to lower frequencies and a new band, clearly an OH, appeared from the dissociation of the adsorbed butene. The bands at 1550 to 1570 cm^{-1} were assigned to a π-allyl species. (The shift of the double-bond stretching for butene was about 100 cm^{-1} compared to the shift of 107 cm^{-1} observed for the π-allyl species formed from propylene.)

When C_3D_6 was adsorbed on ZnO, a new strong band instantaneously appeared at 2653 cm^{-1}, which supports the view that propylene is adsorbed dissociatively to form OD. The surface hydroxyl fragments formed by the dissociative adsorption of deuterium-labelled propylene are summarized in Table II. These results show that the dissociative adsorption of propylene

Table II. Chemisorption of labelled propylene on ZnO and changes in the surface fragments, OH and OD

	Compound	Surface fragments	Spectrum
I	$CH_3-CH-CH_2$	OH	Stable
II	$CD_3-CH=CD_2$	OD	Stable
III	$CH_3-CD=CH_2$	OH	Stable
IV	$CH_3-CH=CD_2$	OH	Changes
V	$CD_3-CH=CH_2$	OD	Changes
VI	$CD_3-CH=CD_2$	OD	Stable

takes place by cleavage of a methyl carbon–hydrogen bond to form an allylic species, and the hydrogen responsible for the OH bands comes from the methyl group.

When $CH_3CH{=}CD_2$ is adsorbed on the surface as shown in Table II, the following surface reactions take place:

$$CH_3{-}CH{=}CD_2 + ZnO \rightleftarrows \underset{-Zn\text{———}}{CH_2{\cdots}CH{\cdots}CD_2}\ \underset{\underset{O-}{|}}{H} \rightleftarrows ZnO + CH_2{=}CH{-}CD_2H$$

$$CH_2{=}CH{-}CD_2H + ZnO \rightleftarrows \underset{-Zn\text{———}}{CH_2{\cdots}CH{\cdots}CDH}\ \underset{\underset{O-}{|}}{D} \rightleftarrows ZnO + CHD{=}CH{-}CH_2D$$

etc.

Immediately after the adsorption, an OH band appears at 3593 cm^{-1} but with time the intensity decreases and an OD band appears at 2653 cm^{-1}. The OH initially formed by the dissociated adsorption of $CH_3{-}CH{=}CD_2$ is gradually replaced by OD to reach an equilibrium ratio of surface OH to OD of 3:2 since the two ends of the molecule become equivalent.

In the C—H stretching and deformation region, the initial spectra of the surface hydrocarbons from $CH_3{-}CH{=}CD_2$ (IV) and $CD_3{-}CH{=}CH_2$ (V) are the same within experimental error, which strongly suggests that propylene is adsorbed to form a symmetric allylic species:

$$CD_3{-}CH{=}CH_2 \xrightarrow{-D} CD_2{\cdots}CH{\cdots}CH_2 \xleftarrow{-H} CD_2{=}CH{-}CH_3$$

or

$$CD_2{=}CH{-}\underset{*}{CH_2} \rightleftarrows \underset{*}{CD_2}{-}CH{=}CH_2$$

Figure 7 demonstrates the changes in the estimated OH and OD concentrations with time.[8] The half-life for the decay of OH from the adsorption of $CH_3CH{=}CD_2$ is about 40 min for first order behaviour, this would give a rate of $2{\cdot}9 \times 10^{10}$ molecules $s^{-1}\,cm^{-2}$. If we assume that the isomerization of the olefin proceeds with each hydrogen exchange, the rate of the surface isomerization of $CH_3CH{=}CD_2$ should be 1·5 times as fast as that of the OH disappearance, that is, $4{\cdot}3 \times 10^{10}$ molecules $s^{-1}\,cm^{-2}$. Taking experimental error into consideration, this is in excellent agreement with the observed catalytic rate, $3{\cdot}7 \times 10^{10}$ molecules $s^{-1}\,cm^{-2}$, which strongly suggests that the π-allyl species observed by the infrared technique is the species responsible for the 1,3-hydrogen shift, i.e. the "isomerization" of propylene.

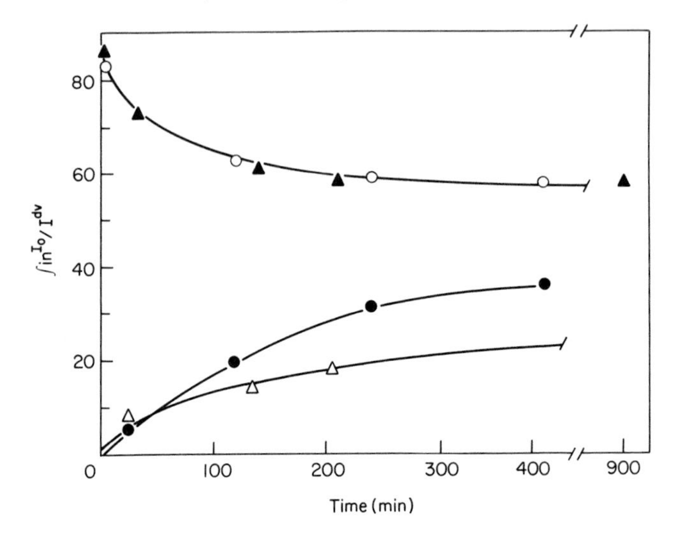

Fig. 7. Integrated intensity of OH and OD bands vs. time for adsorbed labelled propylenes: ○, OH; ●, OD for $CH_3CH{=}CD_2$; △, OH; ▲, OD for $CD_3CH{=}CH_2$.[8]

The rate of the exchange reaction may also be studied by means of mass spectrometry which measures the masses of propylene or hydrogen molecules in the ambient gas. However, mass spectrometry can only determine the overall deuterium content of the molecules, and the deuterium distribution in the molecule, or the deuterium content at each hydrogen position of the propylene, must be determined by another method.

Another technique of following the hydrogen transfer was performed by *Kondo et al.* who used microwave spectroscopy.[11–13] Since microwave spectroscopy has extremely high resolution and high sensitivity, it can be applied to reaction studies to give quantitative analysis of those gaseous molecules with permanent dipole moments. The rotational transition frequencies observed with a microwave spectrometer are dependent on the moments of inertia of the molecules and therefore slight differences in the moments of inertia (usually one part in 10^5) give rise to different spectra.

Propylene-d_1 (C_3H_5D) has four geometrical isomers: propylene-3-d_1 ($CH_2D{-}CH{=}CH_2$), propylene-2-d_1 ($CH_3{-}CD{=}CH_2$), (Z)-propylene-1-d_1 (or cis-propylene-1-d_1),

$$\begin{array}{ccc} CH_3 & & D \\ & \diagdown\;\; \diagup & \\ & C{=}C & \\ & \diagup\;\; \diagdown & \\ H & & H \end{array}$$

, and (E)-propylene-1-d_1 (or

trans-propylene-1-d$_1$), $(CH_3)(H)C{=}C(H)(D)$ [CH$_3$ upper left, H lower left; H upper right, D lower right]. Propylene-d$_2$ ($C_3H_4D_2$) has seven isomers (Table III). The rotational transitions of these isomers are observed at frequencies completely separated from each other. Hence quantitative analysis by microwave spectroscopy can determine the amount of each of the geometrical isomers of propylene-d$_1$ and -d$_2$, even in a mixture of all the possible isomers.

Even if a surface π-allyl species is observed on ZnO, it does not follow that it is a reaction intermediate through which hydrogen exchange or isomerization takes place. If the inter- and intramolecular hydrogen exchange reaction takes

Table III. The geometrical isomers of propylene-d$_1$ and -d$_2$

(Structures written as: upper-left, lower-left substituents on the first C; upper-right, lower-right substituents on the second C.)

Propene-d$_1$ isomers	Propene-d$_2$ isomers	
$(D)(H)C{=}C(CH_3)(H)$ (*Z*)-Propene-1-d$_1$	$(D)(D)C{=}C(CH_3)(H)$ Propene-1,1-d$_2$	$(D)(H)C{=}C(CH_2D)(H)$ (*Z*)-Propene-1,3-d$_2$
$(H)(H)C{=}C(CH_3)(H)$ (*E*)-Propene-1-d$_1$	$(D)(H)C{=}C(CH_2)(D)$ (*Z*)-Propene-1,2-d$_2$	$(H)(D)C{=}C(CH_2D)(H)$ (*E*)-Propene-1,3-d$_2$
$(H)(D)C{=}C(CH_3)(D)$ Propene-2-d$_1$	$(H)(D)C{=}C(CH_3)(D)$ (*E*)-Propene-1,2-d$_2$	$(H)(H)C{=}C(CH_2D)(D)$ Propene-2,3-d$_2$
$(H)(H)C{=}C(CH_2D)(H)$ Propene-3-d$_1$		$(H)(H)C{=}C(CHD_2)(H)$ Propene-3,3-d$_2$

place via the surface π-allyl species, the distribution of the reaction products should be revealed in a characteristic fashion as has been explained elsewhere[12].

When the hydrogen exchange reaction takes place between propylene and a substance containing deuterium, at least one of the hydrogen atoms in the propylene molecule should be dissociated prior to, after, or simultaneously with the deuterium atom addition in a dissociative, associative, or concerted mechanism, respectively. There are two possible intermediates for an associative mechanism, 1-propyl and 2-propyl half-hydrogenated states of the olefin. In the dissociative mechanism there are more possible intermediates: 1-propenyl, 2-propenyl, σ-allyl, and π-allyl intermediates. These seven different mechanisms, including the concerted (push–pull) mechanism, would give different reaction products as demonstrated in Table IV.[12]

If the 1-propyl species is an intermediate, only propylene-2-d_1 should appear as shown in Scheme (a) in Table IV. The 2-propyl intermediate, $CH_2D{-}\overset{*}{C}H{-}CH_3$, formed by the addition of a deuterium atom to propylene, will produce propylene-d_1 isomers by dissociating one of the five hydrogen atoms of the two methyl groups, as shown in Scheme (b). Neglecting the secondary kinetic isotope effect and according to an equal probability of the dissociation of each hydrogen atom of CH_2D— and CH_3—, the d_1 species thus formed should be propylene-3-d_1 (60%) and propylene-1-d_1 (40%). By similar considerations, the distribution of propylene-d_2 isomers should be propylene-1,1-d_2 (10%), propylene-1,3-d_2 (60%), and propylene-3,3-d_2 (30%), as shown in Scheme (b). These initial distributions of propylene-d_1 and -d_2 isomers are the same as those in the equilibrium distribution, since the five hydrogen atoms of the methyl and methylene groups of the propylene take part equally in the exchange reaction.

In the case of Scheme (f), in the initial stages, propylene-3-d_1 and propylene-3,3-d_2 are the only isomers formed via the σ-allyl intermediate and no deuteropropylene containing more than three deuterium atoms should appear from this mechanism.

In the case of the π-allyl intermediate, the end carbons, C-1 and C-3, become equivalent, by dissociation of one of the three hydrogen atoms of the methyl group. Propylene-d_1 isomers are formed by adding a deuterium atom to one of the two methylene groups of the π-allyl intermediate. Consequently, the propylene-d_1 isomer initially formed is solely propylene-3-d_1, deuterium appearing in the methyl group only. However, propylene-1-d_1, propylene-1,3-d_2, propylene-3,3-d_2, and propylene-1,1-d_2 are produced in time by repetition of the dissociation–addition process until an equilibrium isotopic distribution is reached at both end carbons. Since propylene has five hydrogen atoms which participate in this exchange reaction, the statistical equilibrium distribution of propylene-d_1 and -d_2 isomers for this mechanism is the same as that

Table IV. Hydrogen exchange reaction of propylene

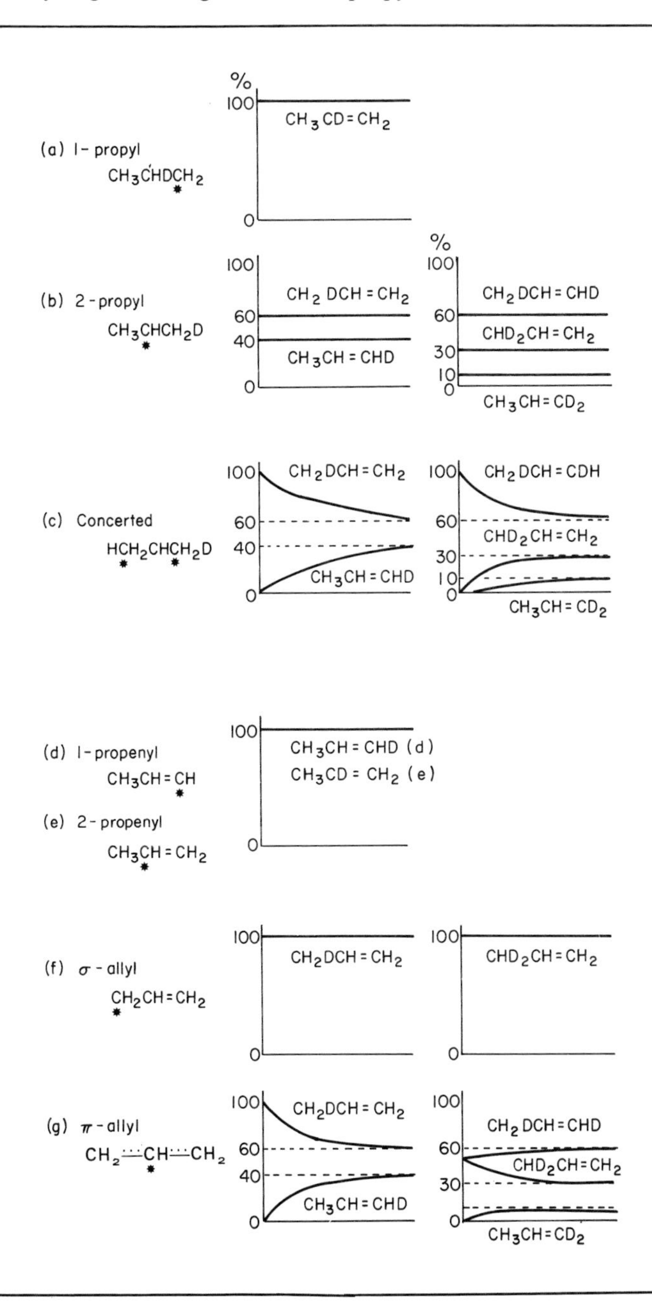

for the 2-propyl intermediate mechanism. In the latter case, however, the distribution stays unchanged throughout the reaction, whereas in the former it changes with time as shown in Table IV.

If a concerted mechanism takes place, one of the hydrogen atoms of the methyl group in the propylene is dissociated, a deuterium atom being simultaneously added to the methylene group. The initial d_1-product is, accordingly, propylene-3-d_1 only, while for the d_2 species it is only propylene-1,3-d_2. The three mechanisms, σ-allyl and π-allyl intermediates and the concerted mechanism, all give propylene-3-d_1 (100%) as the initial exchange product, but they can be distinguished with ease by studying the behaviour of the d_2 species as is shown in Table IV. If more than two mechanisms take place simultaneously, the role of each may be quantitatively estimated by analysing the data on the basis of Table IV. This new approach for elucidating the mechanisms of hydrogen exchange between olefins and deuterium-containing species and of isomerization has been applied to various reaction systems; for instance, the isomerization of propylene on *p*-toluene sulphonic acid, phosphoric acid, or sulphuric acid is shown to proceed via the isopropyl species, accepting protons at the methylene position.[13] It does not proceed through the concerted mechanism which has sometimes been claimed to be the case. The reaction D_2O + propylene on Mo-Bi-oxide proceeds via a σ-allyl intermediate at 100°C[12] and on Al_2O_3 proceeds via a 1-propenyl intermediate forming (Z)-propylene-1-d_1 in the initial stage of the reaction.[14]

A mixture of propylene and deuterium was introduced at room temperature onto ZnO, and the deuterium distribution of the product propylene-d_1 and -d_2 species determined by microwave spectroscopy. Only propylene-3-d_1 was formed, according to Naito, Kondo, Ichikawa and Tamaru, in the initial stage of the reaction, but in the later stages three kinds of monodeuteropropylene, 3-d_1, (Z)-1-d_1, and (E)-1-d_1, were formed in the approximate ratio of 3:1:1. Propylene-2-d_1 was not observed during the reaction. The deuterium distribution in the propylene-d_2 species was observed as shown in Fig. 8. It is obvious from Fig. 8 that propylene-1,3-d_2 and propylene-3,3-d_2 are formed in equal amounts at the beginning of the reaction. This indicates that C-1 and C-3 become equivalent in the reaction intermediate. These experimental results lead to the conclusion that the π-allyl species, which was detected on the catalyst surface by Dent and Kokes, is the real intermediate of the propylene-deuterium exchange and in the double-bond shift isomerization reaction over zinc oxide.

Since Woodman and Taylor[15] reported that the hydrogenation of ethylene proceeds on zinc oxide, a number of reports have appeared. The reaction was also studied by means of infrared spectroscopic techniques by Dent and Kokes[16] and by Naito *et al.*[11] When ethylene is admitted onto ZnO, no bands are observed in the ZnH or OH region, and bands due to adsorbed

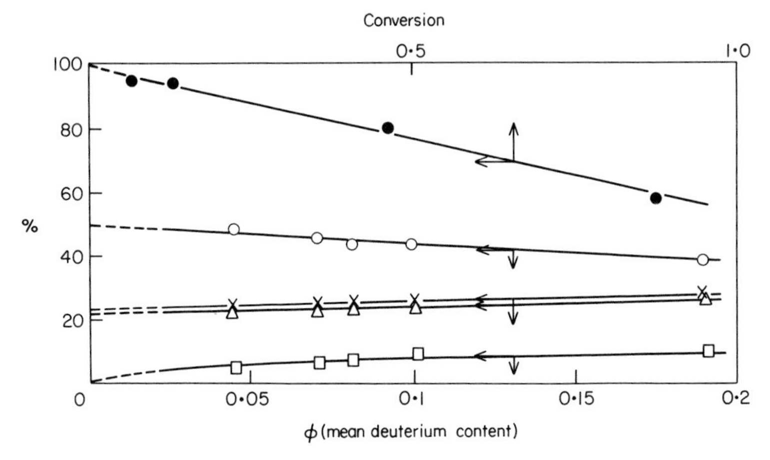

Fig. 8. Hydrogen exchange reaction between propylene and deuterium, studied by means of microwave technique, on ZnO at room temperature: (○)HD_2CCHCH_2, (△)$H_2DCCHCHD$(cis), (×)$H_2DCCHCHD$(trans), (□)H_3CCHCD_2, (●)$H_3CCHDCH_2D$. Conversion = propane/(propylene + propane)

$$\phi = \sum_{i=1}^{6} id_i/6 \sum_{i=0}^{6} d_i.^{(11)}$$

ethylene appear in the C—H region at 3055, 2984, 2993, and 3125 cm^{-1}. Bands at 1600 cm^{-1} are assigned to the C=C stretch of chemisorbed ethylene and bands at 1451 and 1438 cm^{-1} are assigned to CH deformations. This assignment may be supported by the spectrum of perdeuteroethylene, which gives only one band in the region at 1495 cm^{-1}. Dent and Kokes interpreted these spectra to suggest that the chemisorbed ethylene is "π-bonded" to the surface.[10]

When gaseous deuterium is mixed with ethylene and the hydrogenation of ethylene starts, the band at 2984 cm^{-1} first broadens and then decreases in intensity as new bands appear at 2892, 2860, and 2812 cm^{-1}. In the deformation region the two ethylene bands at 1451 and 1438 cm^{-1}, appear to weaken and to shift slightly. In addition, a new band appears at about 1415 cm^{-1}. Because the bands associated with chemisorbed ethylene decrease in intensity on the addition of deuterium, it seems that under the reaction conditions a major portion of the chemisorbed ethylene is converted to a new species. If the ambient ethylene gas is trapped out, (it would have been better to have replaced it by C_2D_4 in order to keep the surface conditions unchanged) and circulation is continued, the new bands decrease initially at a rapid rate and then decrease more slowly. The initial rate of removal of this species is comparable to, or somewhat less than, the rate of the steady-state catalytic hydrogenation, thus suggesting the possibility that the species is the reaction intermediate.

Dent and Kokes interpreted the above behaviour as

$$C_2H_4 + 2H(a) \rightarrow C_2H_5 + H(a) \rightarrow C_2H_6(g)$$

and the decay of the intensities of the intermediate bands represents the reaction between the ethyl groups and dissociatively adsorbed hydrogen. Because the reaction of ethylene with deuterium yields $C_2H_4D_2$ only, then if hydrogenation takes place via adsorbed ethyl groups, the formation of the ethyl intermediate from the adsorbed ethylene cannot be reversible. A reaction order of 0·5 with respect to deuterium and a slight dependence upon ethylene pressure are obtained, which is different from the results for higher olefins and for butadiene.

The mechanism of deuterium addition to propylene, butene, and 1,3-butadiene was also studied by Naito, Sakurai, Shimizu, Onishi and Tamaru over ZnO at room temperature.[17]

The addition of deuterium to 1-butene proceeded over ZnO at room temperature, and isomerization took place simultaneously. The isomerization also occurred in the absence of deuterium and a comparison of the initial rates in these two cases is given in Table V. The rate of isomerization was not affected by the presence or absence of deuterium in the ambient gas. The deuterium distribution was studied for the reaction between 1-butene and D_2 (Fig. 9) over ZnO and butane-d_2 was found to be the only product in the initial stages and hydrogen–deuterium exchange within the 1-butene was very small. In addition, in the isomerization process, 2-butene was almost normal;

Table V. Comparison of the initial rate of butene isomerization in the presence and absence of deuterium gas, over ZnO at room temperature

	Initial rates of the formation $cm^3\ min^{-1}$			
Initial reactants	1-butene	c-2-butene	t-2-butene	butane
1-butene + D_2		0·29	0·09	0·11
1-butene		0·28	0·10	
c-2-butene + D_2	0·02		0·08	0·01
c-2-butene	0·03		0·09	
t-2-butene + D_2	0·004	0·09		0·01
t-2-butene	0·003	0·09		

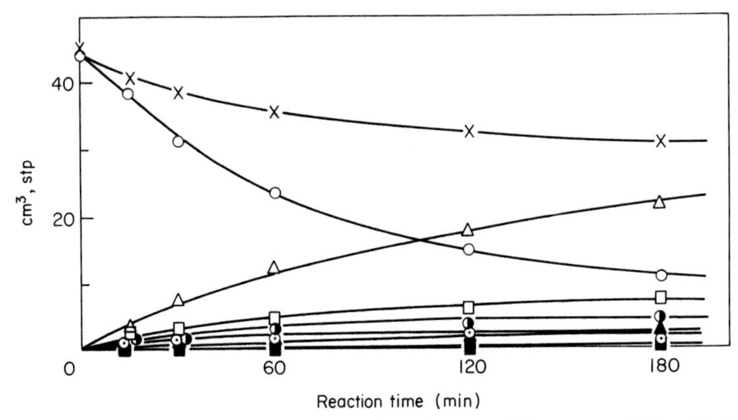

Fig. 9. Deuterium distribution in the reaction between 1-butene and D_2; ZnO, 18 g; temp. 25°C. (○), 1-butene-d_0 (C_4H_8); (▲), 2-butene-d_1 (C_4H_7D); (⊙), 1-butene-d_1 (C_4H_7D); (□), butane-d_2 ($C_4H_8D_2$); (△), 2-butene-d_0 (C_4H_8); (■), butane-d_1 (C_4H_9D); (×) D_2; (◑), HD.[17]

deuterium did not enter cis-2-butene or trans-2-butene at the beginning of the reaction. These results suggest that both deuteration and isomerization proceed without interaction with chemisorbed hydrogen. When a mixture of 1-butene, H_2 and D_2 was introduced over ZnO, the main products were initially butane-d_0 and butane-d_2, but the H_2–D_2 exchange reaction was rapid so that the concentration of butene-d_1 increased as the reaction proceeded.

Deuterium addition to 1,3-butadiene over ZnO occurred readily at room temperature, 1-butene being the main product (>90%) until all of the 1,3-butadiene in the gas phase was consumed, and then butane was produced from 1-butene, accompanied by the isomerization described above. Butene-d_2 (100%) was the only butene produced during the reaction and according to microwave spectroscopic measurements it had the structure $H_2DC{-}CHD{-}CH{=}CH_2$. On the other hand, the hydrogen atoms of 1,3-butadiene never exchanged with deuterium molecules in the gas phase.

When a mixture of 1,3-butadiene + H_2 + D_2 (1:1:1) was introduced over ZnO, the main products were 1-butene-d_0 and 1-butene-d_2, while the amount of 1-butene-d_1 was very small and proportional to the amount of HD in the gas phase, as is shown in Table VI. The last column of Table VI gives the rate of the H_2–D_2 (1: 1) exchange reaction in the absence of 1,3-butadiene. The H_2–D_2 exchange reaction occurred to a small extent during the deuteration reaction but it was strongly inhibited by the presence of 1,3-butadiene. These results indicate that either deuterium participates in the addition process in molecular form or that the half-hydrogenated state is formed irreversibly.

Table VI. Deuterium contents in the main products produced by the reaction of $C_4H_6 + H_2 + D_2$ (1:1:1) over ZnO at 29°C

reaction time (min.)	conv. (%)	C_4H_6	1-butene C_4H_2	C_4H_7D	$C_4H_6D_2$	hydrogen H_2	HD	D_2	H_2–D_2 exchange† H_2	HD	D_2
10	6·9	1·00	0·46	0·08	0·46	0·45	0·07	0·48	0·29	0·40	0·31
20	10·4	1·00	0·51	0·11	0·38	0·42	0·01	0·48	0·26	0·46	0·27
40	16·9	1·00	0·48	0·17	0·35	0·31	0·18	0·51	0·25	0·49	0·26
80	27·0	1·00	0·46	0·20	0·34	0·33	0·19	0·48			
160	45·0	1·00	0·44	0·26	0·30	0·34	0·26	0·40			

† H_2–D_2 exchange reaction in the absence of 1,3-butadiene under similar conditions.

Naito *et al.*[17] also examined the adsorption during the course of the reactions, simultaneously measuring overall reaction rates. The deuterium addition was found to be first order with respect to the deuterium pressure and zero order to the olefins or 1,3-butadiene. The amounts of hydrogen and hydrocarbons chemisorbed during the hydrogenation, as determined volumetrically, remained unchanged throughout the reaction as shown in the

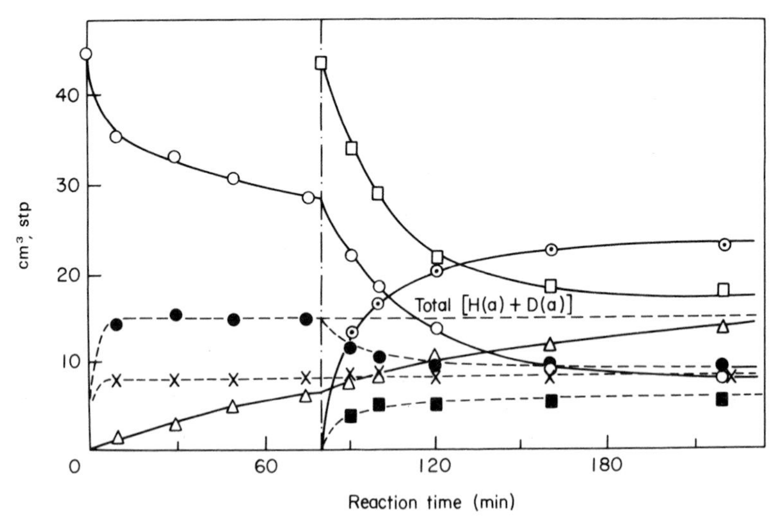

Fig. 10. Adsorption measurements during the hydrogenation of butene: (1-butene + H_2) + D_2 over ZnO at 240°C. (○) H_2(g); (●) H(a); (□) D_2(g); (■) D(a); (⊙) HD(g); (×) 1-butene(a); (△) butane(g).[17]

example in Fig. 10. The ratio of adsorbed H_2 to adsorbed hydrocarbon was approximately 4: 1 for propylene, 1: 1 for butene, and 1: 4 for butadiene, i.e. the amount of hydrogen chemisorbed at saturation decreased in the order: propylene > butene > butadiene. The H_2–D_2 exchange reaction during hydrogenation was retarded in the same order: propylene (half-life = 6 min) > butene (14 min) > butadiene (73 min).

When hydrogen is admitted over ZnO at room temperature, there are two types of hydrogen chemisorbed, a fast (H_{rev}) and a slow (H_{irrev}) chemisorption. The former is easily removed from the surface by evacuation and is the reaction intermediate of the H_2–D_2 exchange reaction. H_{irrev}, however, which is a minor part of the total hydrogen chemisorbed, is only slowly removed at room temperature. The rate of hydrogenation of olefins by H_{irrev} is very slow, whereas the reaction between H_{rev} or H_2 and olefins proceeds at a considerable rate at the same temperature, and its rate is not affected by the presence of H_{irrev}.

The rate of hydrogenation of butene, for example, is independent of butene pressure, the amount of adsorbed butene remaining unchanged irrespective of the olefin pressure. On the other hand, the rate is proportional to the hydrogen pressure. The amount of hydrogen chemisorbed on the catalyst surface during the reaction remains constant throughout the reaction and is independent of the hydrogen pressure. Consequently, the rate of butene hydrogenation is dependent upon the partial pressure of hydrogen and not upon the amount of hydrogen chemisorbed on the catalyst surface.

If hydrogenation of butene over ZnO proceeds via a half-hydrogenated state, the hydrogen participating in the reaction cannot be that measured in the volumetric adsorption measurements, as this remains constant even though the ambient hydrogen pressure and rate of reaction are increased. The formation of the half-hydrogenated state must be irreversible and the participating hydrogen must be proportional to the ambient hydrogen pressure. In this sense, the participating hydrogen can be said to be in a "molecular form".

4-4 Catalytic decomposition of formic acid

Substances such as alumina and silica are good catalysts for dehydration reactions whereas over zinc oxide and magnesium oxide the main reaction is generally dehydrogenation. Often both reactions occur simultaneously and the catalytic selectivity, i.e. the extent to which one or other reaction predominates is in general dependent on the method of preparation of the catalyst. The dehydration/dehydrogenation of formic acid has been studied by many workers as a model of catalyst selectivity.[18]

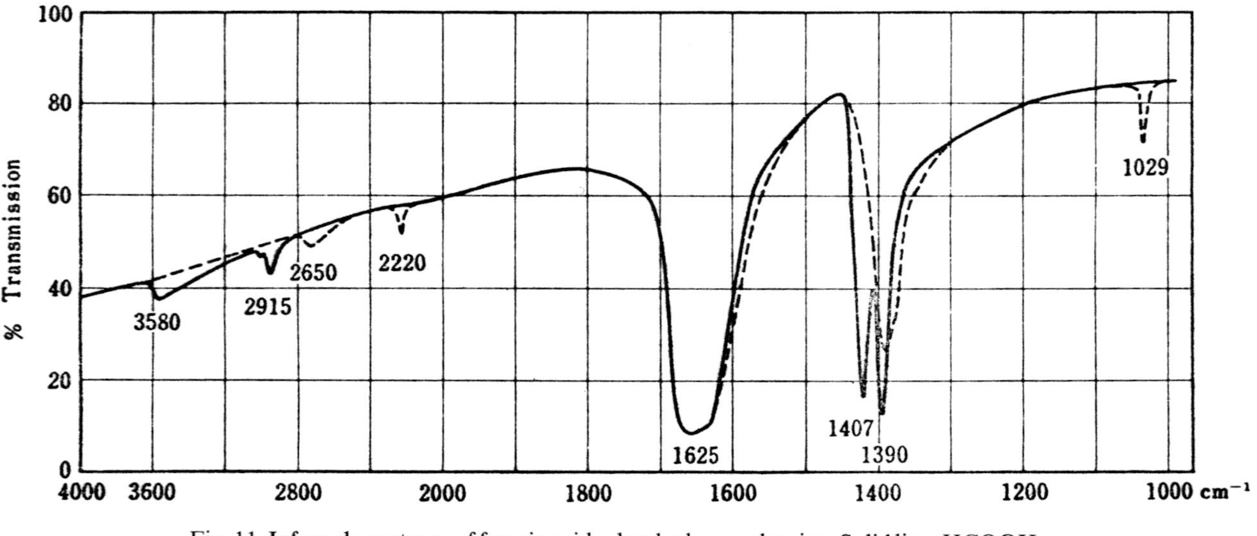

Fig. 11. Infrared spectrum of formic acid adsorbed on γ-alumina. Solid line, HCOOH; broken line, DCOOD.[19]

During the decomposition of formic acid on various metals and oxides, it has been reported that surface formate ions can be observed by infrared spectroscopy, and accordingly it was, in many cases, concluded that the decomposition of formic acid proceeds via this species. However, the fact that the formate species exists on the catalyst surface during the reaction, does not necessarily mean that it is a reaction intermediate. Tamaru and his co-workers have studied the decomposition by means of their "dynamic treatment".[19–21]

When alumina is exposed to formic acid vapour, formate ions and surface OH groups are formed on the alumina by dissociative adsorption: a spectrum in the range 4000 to 1000 cm^{-1} of HCOOH adsorbed on alumina at 100°C (see Fig. 11) is very similar to that of aluminium formate. The bands at 2915, 1625, 1407, and 1390 cm^{-1} can be assigned to the CH stretching, O—C—O antisymmetric stretching, CH in-plane bending, and O—C—O symmetric stretching vibrations, respectively. In the spectrum of DCOOD adsorbed on alumina, the CH stretching and CH in-plane bending vibrations are shifted to lower frequencies, 2220 and 1029 cm^{-1}, respectively, supporting the assignments.

The surface formate ion is stable and *in vacuo* decomposes to mainly H_2O and CO only at temperatures considerably higher than those at which formic acid decomposition takes place. The decomposition rate of the formate species on the alumina catalyst *in vacuo* is about two orders of magnitude slower than the rate of decomposition of formic acid vapour at the same reaction temperature and coverage.

A known amount of DCOOD was introduced into the reaction system and the rate of the overall dehydration decomposition was measured by the change in pressure and composition of the gas phase. The adsorption on the catalyst was followed simultaneously by means of infrared spectroscopy.

When the reaction had reached a steady state, the DCOOD in the gas phase was quickly replaced by HCOOH and the reaction continued. The behaviour of the chemisorbed species as well as the overall reaction rate were followed simultaneously. The amounts of the formate ions, $DCOO^-$(a) and $HCOO^-$(a), were estimated from the C–D and C–H stretching vibrations, respectively. (Typical infrared spectra are shown in Fig. 12.)

The results showed that some of the adsorbed formate ions ($DCOO^-$) desorb directly into the gas phase as formic acid vapour, without decomposing into the reaction products. Some ambiguity is involved in the conclusions, however, due to the exchange reaction and the scatter in the experimental results.

When alumina was pretreated with HCl or acetic acid vapour, the rate of formic acid decomposition was not influenced. The decomposition of adsorbed acetic acid did not take place to an appreciable extent, and there was no detectable exchange between the acetate and formate ions. The amount of

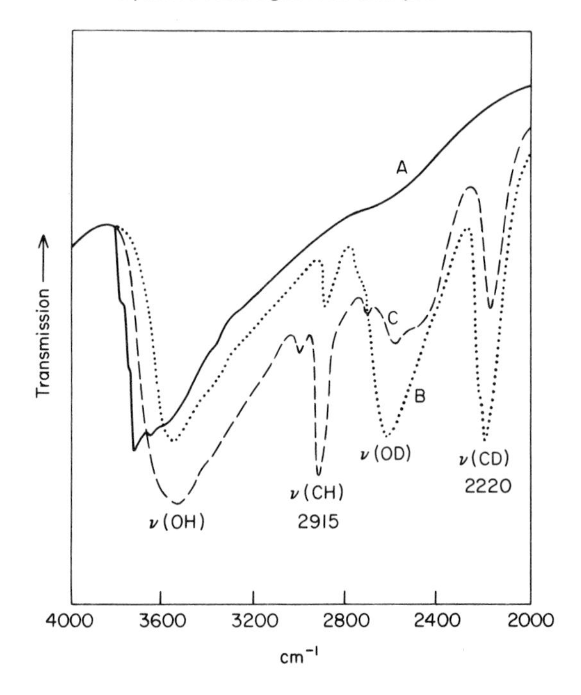

Fig. 12. Decomposition of formic acid on γ-alumina by the dynamic treatment with the infrared technique (at 190°C): (A) background spectrum of γ-alumina, (B) DCOOD adsorbed, (C) reaction with HCOOH.[19]

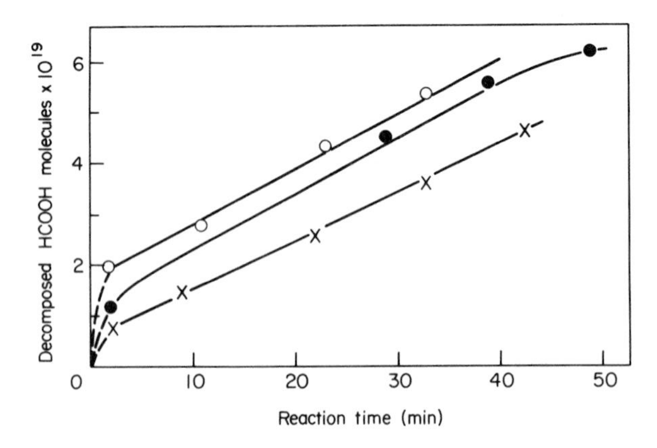

Fig. 13. Decomposition of formic acid on an alumina pretreated by acetic acid. (×), pretreated alumina; (○, ●), without pretreatment.[19]

formate ion on the alumina was decreased by the acetic acid pretreatment to about one-tenth of its normal value, but the rate of formic acid decomposition was approximately the same in both cases as shown in Fig. 13. This is another strong indication that the activity of the formate ion adsorbed on the alumina does not determine the decomposition rate.

The next experiment undertaken was the direct volumetric measurement of adsorption while the reaction was taking place. It was demonstrated by infrared studies that during the formic acid decomposition on alumina the chemisorbed species consist of formate ions, protons (OH), and water. There were no carbon monoxide, carbon dioxide, or formic acid molecules adsorbed under reaction conditions. The amount of adsorption of each species during the course of the reaction was estimated from the material balance of C, H, and O in the closed circulating system. The partial pressures of the reactant and products in the gas phase were also measured.

In the steady state of the overall reaction, the rate could be expressed as

$$r = kP_{\mathrm{HCOOH}}(H^+)/(1 + b(\mathrm{H_2O})_{\mathrm{ads}})$$

The form of the rate equation suggests that the reaction takes place between surface protons and formic acid molecules in the ambient gas. These results accordingly lead to the conclusion that the dissociative adsorption of formic acid supplies protons on the catalyst surface which behave as reaction sites for subsequent reaction. The formate ion on the catalyst surface does not behave as the main reaction intermediate.

This conclusion does not exclude the possibility that the decomposition takes place via the formate ion under other reaction conditions. As the decomposition via the formate ion seemingly has a higher activation energy than the decomposition by the interaction of formic acid with the surface protons, the decomposition may proceed through the formate ion at much higher temperatures.

The mechanism of dehydrogenation decomposition of formic acid over zinc oxide was studied in a similar manner.[20] In this case, the formic acid absorbed into the bulk of the oxide and a large quantity of water appeared in the gas phase. That the formic acid had reacted to form the metal formate and water was confirmed by infrared spectra.

The decomposition of formic acid was studied over zinc formate, and it was found that the reaction proceeded in the same temperature range and with the same reaction order as over the oxide. It was accordingly confirmed that the decomposition of formic acid on zinc oxide actually proceeds over the formate.

Kinetic studies of the decomposition on ZnO were carried out in the temperature range 145—220°C in which range the decomposition was more than 95% dehydrogenation. The rate of the decomposition was independent of the pressure of the formic acid. Hydrogen and carbon dioxide were produced at the same rate provided formic acid was present in the gas phase. However, it is very interesting to note that if all the formic acid was trapped out of the ambient gas or was consumed, the evolution of hydrogen stopped while the rate of carbon dioxide evolution remained constant. When formic acid was re-introduced into the gas phase, hydrogen evolved rapidly until the amount of carbon dioxide and hydrogen became the same, as shown in Fig. 14.

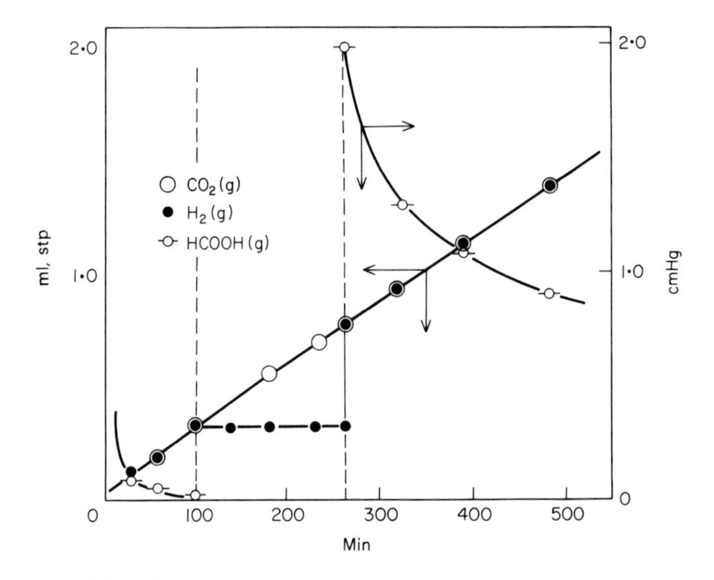

Fig. 14. Dehydrogenation of formic acid on ZnO.[20]

It is of interest to note that when HCOOD was decomposed over ordinary zinc formate, the hydrogen produced was practically all HD. At these decomposition temperatures the exchange reaction between D_2 and H_2 was negligible over the catalyst.

On the basis of these results, the mechanism of the dehydrogenation of formic acid over zinc oxide or formate is shown to be

$$\underset{-\mathrm{Zn}-}{\mathrm{HC(O)O}} \xrightarrow{A} \overset{CO_2\uparrow}{\underset{-\mathrm{Zn}-}{\mathrm{H}}} \xrightarrow{\mathrm{HCOOD}} \underset{-\mathrm{Zn}-}{\mathrm{H}} \quad \mathrm{DO{-}C(H){=}O} \longrightarrow \overset{\mathrm{HD}\uparrow}{\underset{-\mathrm{Zn}-}{\mathrm{HC(O)O}}}$$

The evolution of carbon dioxide arises from the decomposition of formate ions, the hydrogen atoms remaining on the surface in the absence of formic acid in the ambient gas. Gaseous hydrogen appears when formic acid molecules attack the surface hydrogenation atoms, producing formate ions on the surface. In confirmation of this, only HD appeared when HCOOD was decomposed on zinc formate. In this case the hydrogen molecule is always produced from the hydrogen atom of the OH group in the formic acid molecule and the surface hydrogen atoms.

The rate-determining step is the decomposition of the formate ions on the surface. Because the formate ion saturates the surface in the presence of formic acid, the rates of evolution of CO_2 and H_2 are independent of the pressure of the formic acid. The decomposition of formic acid on various metal surfaces, in particular, nickel, silver and copper has also been studied in a similar manner.

4-5 The water–gas shift reaction on zinc oxide and magnesium oxide

When a mixture of carbon monoxide and water vapour was introduced over magnesium oxide, Scholten and his co-workers observed, by means of infrared spectroscopy, the formation of a surface formate ion.[22] To confirm that the formate ion is the real reaction intermediate of the water–gas shift reaction, the techniques outlined previously were used to measure not only the overall reaction rate, but also the kinetic behaviour[23] of the adsorbed species during the course of the reaction. The coverage of the surface species during the reaction and their kinetic behaviour were measured by infrared spectroscopy in the range 4000 to 900 cm^{-1}. The selectivity of the decomposition of the surface species and the rate of the overall reaction were measured using gas chromatography.

Zinc oxide catalyst was pretreated under vacuum for 5 hours at 400°C. After 2 hours at 150°C in contact with an equimolar mixture (200 Torr) of carbon dioxide and hydrogen, the gas phase was pumped off and the observed infrared spectra of the adsorbed species are shown in Fig. 15.

When 0·1 Torr of formic acid was introduced over the zinc oxide at room temperature, the adsorbed species gave a similar infrared spectrum (Fig. 15), confirming that surface formate ions are produced from a mixture of CO_2 and H_2. The 2870, 1572, and 1369 cm^{-1} bands are attributed to CH stretching, O—C—O antisymmetric stretching, and O—C—O symmetric stretching vibrations, respectively, while the 1379 cm^{-1} band is the C—H in-plane bending vibration. When deuterium was employed instead of hydrogen, the isotope shift of the C—H (or C—D) stretching vibration bands was observed,

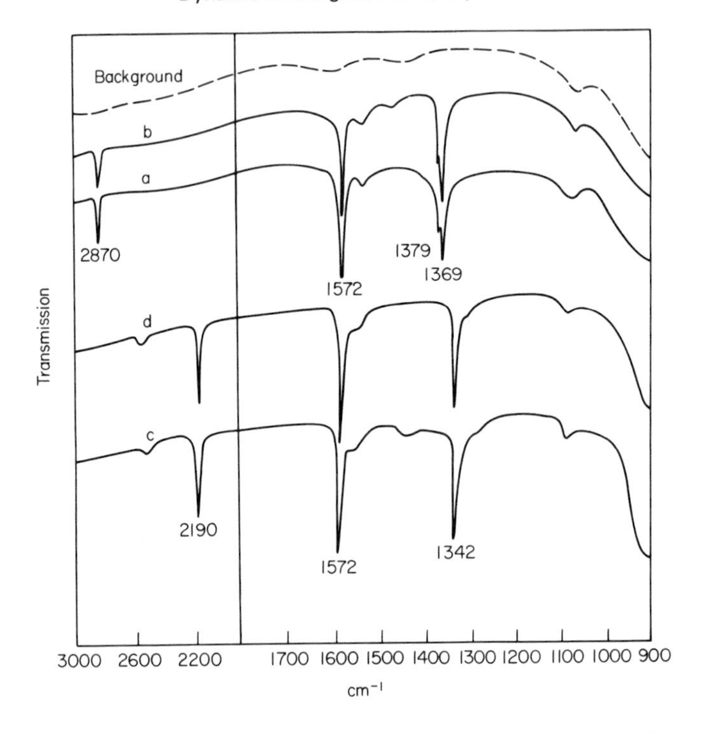

Fig. 15. Infrared spectra of the chemisorbed species on ZnO at 150°C: (a) CO_2+ H_2; (b) HCOOH; (c) CO_2+ D_2; (d) DCOOD.[23]

as shown in Fig. 15. The formation of formate ions was not detected when a mixture of carbon monoxide and water vapour was introduced over the ZnO at 200°C.

The role of the surface formate ion in the water—gas shift reaction was then studied by following its dynamic behaviour at varying surface coverages and temperatures by means of infrared spectroscopy. The formate ion on ZnO is comparatively stable at 200°C, but at 300°C carbon monoxide is formed at a substantial rate by dehydration decomposition.

The rate of the overall reaction was compared with that of formate ion decomposition at the reaction temperature (230°C) by means of infrared spectroscopy. First, a known amount of DCOOD was adsorbed onto the ZnO and the optical density of the C–D stretching vibration band was followed with time. The rate of decomposition of the surface formate ion on ZnO was thus measured at various surface coverages, and the dependence of the rate upon the coverage was obtained.

As water vapour is a product of the reaction between H_2 and CO_2, its effect upon the decomposition rate of the surface formate ion was examined. In Fig. 16, 0·3 cm^3(stp) of water vapour was added to 0·25 g of catalyst during the course of the decomposition of surface formate ions. It may be seen from the figure that the rate of decomposition was not greatly influenced by the adsorption of additional water under the reaction conditions.

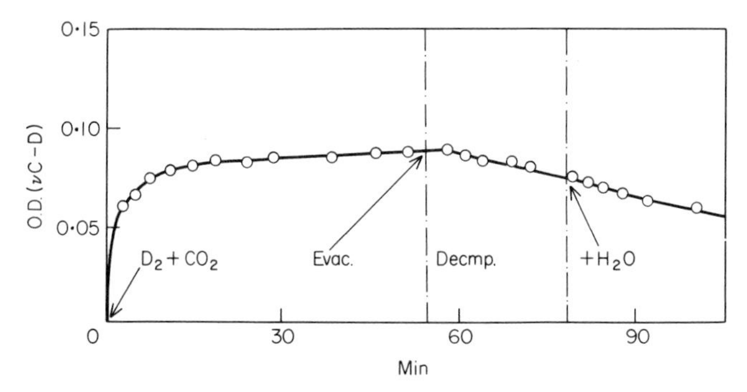

Fig. 16. Influence of the additional water vapour upon the rate of decomposition of surface formate on ZnO.[23]

The disappearance of the surface formate results in the formation of not only carbon monoxide, but also of carbon dioxide. The dependence of the selectivity of CO formation, $CO/(CO+ CO_2)$, upon the coverage of the formate ion at 230°C was also studied, and the rate of CO formation from the surface formate ion decomposition was estimated by multiplying the rate of decomposition of the formate ions by the selectivity at the same coverage.

In addition, the rate of CO formation from the overall reaction and the coverage of surface formate ions were measured by the infrared technique, thus yielding the amounts of CO produced. Consequently, the rate of CO formation by surface formate decomposition was compared with that of the

Table VII. Comparison of the rate of the dehydration of surface formate with that of the overall reaction on ZnO at 230°C

$D_2 + CO_2 \rightarrow D_2O + CO$ ml/g hr	$DCO_2(a) \rightarrow D_2O + CO$ ml/g hr	Coverage of $DCO_2(a)$
0·023	0·022	0·35
0·028	0·025	0·38
0·030	0·029	0·41

overall reaction under a steady-state condition at the same coverage of the formate ion on ZnO, as given in Table VII. The rates are in reasonable agreement within experimental error, leading to the conclusion that the surface formate ion is the real reaction intermediate through which the water–gas shift reaction takes place on ZnO. That is,

$$\left.\begin{array}{l} H_2(g) \rightleftharpoons 2H(a) \\ CO_2(g) \rightleftharpoons CO_2(a) \end{array}\right\} \overset{(1)}{\rightleftharpoons} HCOO(a) + H(a) \overset{(2)}{\rightleftharpoons} \left\{\begin{array}{l} OH(a) + H(a) \rightleftharpoons H_2O(g) \\ CO(a) \rightleftharpoons CO(g) \end{array}\right.$$

with the dehydration step (2), the formation of CO from the surface formate ion, being the rate-determining step of the overall water–gas shift reaction.

With magnesium oxide, when a mixture of carbon monoxide and water vapour was introduced, infrared spectroscopy showed the presence of surface formate ions on the catalyst in agreement with the results of Scholten and his co-workers. In contrast to the ZnO situation, however, surface formate ions were not readily formed from a mixture of carbon dioxide and hydrogen. By means of a dynamic treatment of the surface formate ion similar to the ZnO investigation, it was concluded that the surface formate ion is the intermediate of the reaction on MgO, the rate-determining step being the dehydrogenation step (1) in the scheme described above.

It has therefore been demonstrated by the dynamic treatment of the chemisorbed species, that the surface formate ion is the reaction intermediate of the water–gas shift reaction on ZnO and MgO. These reaction systems are typical in that adsorption studies of the individual reactants on the catalyst surface do not reveal the real reaction intermediate. Only by a study of the behaviour of the reacting mixture under the reaction conditions is it possible to elucidate the mechanism of the reaction.

4-6 Decomposition of methyl alcohol on zinc oxide

The decomposition of methyl alcohol can proceed by a variety of reaction paths, it being capable of producing carbon monoxide, hydrogen, ether, and also carbon dioxide. It is therefore of great interest to determine the reaction mechanism and to identify the real reaction intermediate. Little is known about the catalytic reaction path of the decomposition, although there are many speculative reports about the mechanism of the reaction from the overall kinetic results, and separate adsorption measurements of the reacting gases.[24]

The concentrations and dynamic behaviour of the chemisorbed species and ambient gas molecules were therefore measured, along with the overall

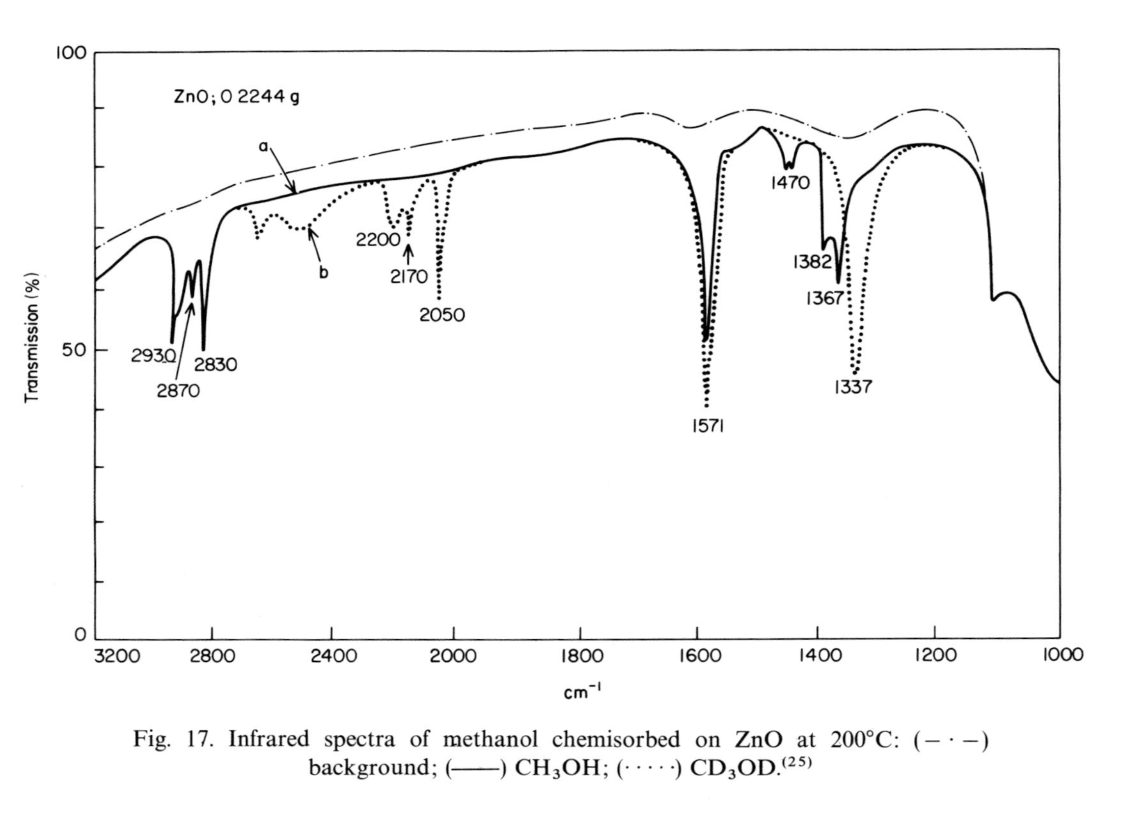

Fig. 17. Infrared spectra of methanol chemisorbed on ZnO at 200°C: (– · –) background; (——) CH_3OH; (· · · ·) CD_3OD.[25]

reaction rate under various non-steady-state conditions of methyl alcohol decomposition over zinc oxide.[25]

Methyl alcohol vapour was introduced over the catalyst, which had been outgassed under vacuum at 400°C for 3 hours, and the infrared spectra of the species adsorbed at 200°C on the zinc oxide were studied. These were obtained by trapping the methyl alcohol from the gas phase and cooling the cell to room temperature. The spectra, shown in Fig. 17, demonstrated the formation of methoxide and formate ions on the surface at 200°C. The bands at 2930, 2830, and 1470 cm^{-1} are assigned respectively to the CH_3 assymetric and symmetric stretching vibrations and the deformation band of the methyl group in the surface methoxide ions. The bands at 2870, 1571, and 1367 cm^{-1} are assigned to the C—H stretching, O—C—O antisymmetric stretching, and O—C—O symmetric stretching vibrations of surface formate ions, respectively. The C—H in-plane bending vibration band of the formate was also observed at 1382 cm^{-1}. When CD_3OD was used, the isotope shift of the C—H(C—D) stretching vibration bands was observed, as shown in Fig. 17.

The optical density of the CD_3 stretching band at 2050 cm^{-1} was measured at 100°C by introducing known amounts of methyl alcohol (CD_3OD) onto the zinc oxide. The dependence of optical density upon coverage was determined for the methoxide ion and, in a similar manner, for the formate ion. Thus the amounts of the surface formate and methoxide on the catalyst could be estimated during the course of the reaction from their optical densities.

The dynamic behaviour of the chemisorbed species is shown in Figs 18 and 19. When methyl alcohol was introduced onto the zinc oxide catalyst at 200°C, hydrogen, carbon dioxide, and carbon monoxide were formed. When the

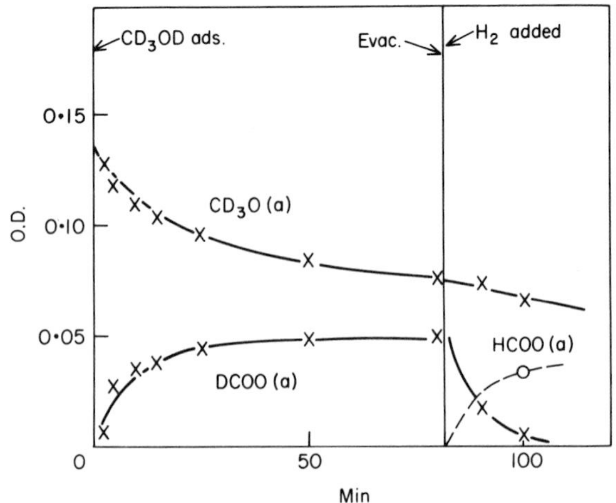

Fig. 18. Surface reaction from methoxide to formate species at 200°C.[25]

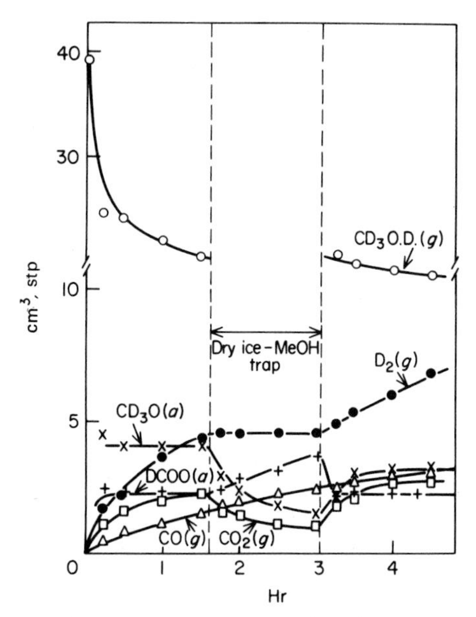

Fig. 19. Adsorption and pressure changes during the course of methanol decomposition on ZnO.[25]

alcohol in the ambient gas was removed from the system by a dry-ice trap, the evolution of hydrogen and carbon dioxide stopped, and the pressure of carbon dioxide gradually decreased. The amount of surface methoxide ion on the catalyst decreased with time, while that of the formate ion increased during the trapping and the evolution of carbon monoxide continued without appreciable change. The sum of the amounts of carbon monoxide produced and surface formate ion formed during the trapping was approximately equal to the decrease in the amount of surface methoxide ion. This suggests the reaction path

$$CH_3O(a) \rightarrow HCOO(a) \rightarrow CO$$

When liquid nitrogen was employed for the trapping instead of dry-ice, thus removing carbon dioxide as well as methanol from the gas phase, the rate of evolution of carbon monoxide was not appreciably influenced. Consequently, the possible reaction path from CO_2 to CO was not considered to be of importance. The main path for the formation of carbon monoxide appears to be via the decomposition of surface formate ions.

The rate of CO formation from the surface formate ion was compared with that from the overall reaction. After removal of the ambient gas, the decomposition rate of the surface formate at 240°C obtained from the infrared spectra was in reasonable agreement with the rate of carbon monoxide production. It was therefore concluded that carbon monoxide is produced by the decomposition of the surface formate ion rather than of the methoxide ion, and the surface formate ion is the only reaction intermediate for CO formation.

In the decomposition of methyl alcohol, the surface formate ion decomposes solely to CO and not to CO_2.

When the trapped methyl alcohol was released (Fig. 19), hydrogen and carbon dioxide were both produced. The concentration of the surface formate ion decreased, whereas that of the methoxide group increased. This behaviour of the surface species and the ambient gases under the reaction conditions suggests that the formation of hydrogen and carbon dioxide proceeds

$$CH_3OH + HCOO(a) \rightarrow H_2 + CO_2 + CH_3O(a)$$

Summarizing all the results obtained by the non-steady-state treatment under the reaction conditions, a mechanism is proposed for the decomposition of methyl alcohol over zinc oxide:

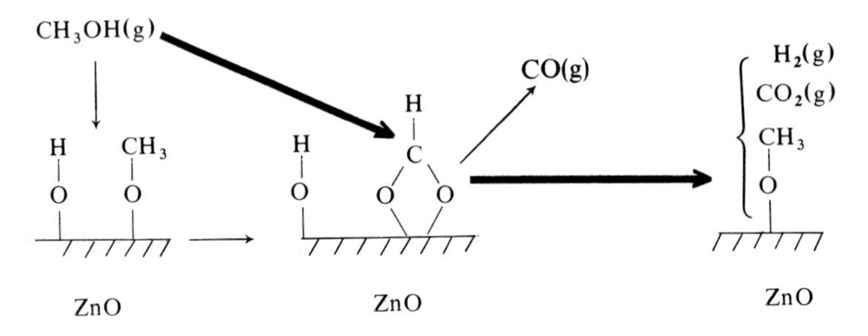

Methyl alcohol is first dissociatively adsorbed on the catalyst surface to form methoxide ions, which then combine with lattice oxygens to form formate ions. These then decompose to form carbon monoxide. The surface formate ion and OH species also react with methyl alcohol to form hydrogen and carbon dioxide, leaving surface methoxide ions.

Because of this successful application of quantitative infrared spectroscopy, it was felt that it would be profitable to extend this work to a detailed study of the individual reaction steps of a similar system. The oxidation of methanol was chosen by Herd *et al.* because the presence of oxygen decreases the infrared absorption by zinc oxide.[26]

The spectra observed for the methoxide, hydroxide and formate surface species on zinc oxide in the presence of oxygen, were the same as those observed on reduced zinc oxide, with the addition of a methoxide band overlapping the ZnO band at 1070 cm^{-1}, which is assigned to C—O stretching. Methoxide groups are strongly bound to the zinc oxide, and at the temperatures studied (140 to 180°C), the rate of desorption was small, at lower coverages.

It is of significant interest to note that methanol oxidation proceeds in the presence of oxygen in a manner similar to that of methanol decomposition in the absence of oxygen, with an additional step, the oxidation of the carbon monoxide and hydrogen formed.[26]

As a key step in the oxidation of methanol, the decomposition of adsorbed methoxide groups was studied in the presence of oxygen. The methoxide groups decomposed to formate ions until a maximum surface concentration was reached. The rate of decomposition at first increased with methoxide concentration, then decreased as the surface coverage approached saturation. The observed kinetics can be explained by a mechanism which postulates that negatively charged surface species repel electrons from the surface of the zinc oxide, thus inhibiting decomposition of the methoxide groups.

The mechanism of heterogeneous catalytic reactions can thus be elucidated by following the dynamic behaviour of the chemisorbed species and ambient gas molecules under the reaction conditions.

4-7 The catalytic oxidation of carbon monoxide on palladium

The catalytic oxidation of carbon monoxide on Pt and Pd is a typical example of a catalytic reaction that has been studied extensively by many investigators.[27–35] The reaction on Pt is also of historical interest because it was the foundation upon which Langmuir successfully pioneered the kinetic study of heterogeneous catalysis.[27] According to the results reported so far, the mechanism of the reaction appears to depend on the experimental conditions and both the Langmuir–Hinshelwood and the Eley–Rideal mechanisms have been suggested.[27–37]

The chemisorption of carbon monoxide and oxygen on the well-defined single crystal faces of Pt and Pd has been examined by means of low-energy electron diffraction, Auger electron spectroscopy and by work function measurements.[31,32,36] These methods have revealed information about the geometrical configuration of the adsorbed species on the surface, the cleanliness of the bare substrate surface and the surface dipoles of the adsorbed species, respectively.

CO oxidation on Pd has been studied extensively by Ertl and his coworkers.[32,34,37] The sticking coefficient for the CO adsorption is high and the LEED pattern reveals that on a (111) surface at about 2/3 of the maximum coverage (ratio Pd:CO = 2:1) there is an orderly arrangement of binding CO molecules in a $\sqrt{3}$ structure. On a (100) face Park and Madden observed a

$2 \times 4/45°$ adsorbed CO structure which also contained bridge bond molecules.[33] A 5×2 structure was observed on a (110) face, and, at higher pressures ($>10^{-6}$ Torr), a 2×1 structure was reported. The latter can be interpreted as being formed by bridge bond CO molecules.

Oxygen, on the other hand, is chemisorbed more strongly on Pd, giving a 2×2 structure on the (111) and (100) faces. The heats of adsorption are of the order of 30 kcal/mol for CO and 55 kcal/mol for oxygen on all faces. When CO is introduced onto an oxygen-covered surface, CO_2 appears readily in the gas phase even at room temperature. In other words, the reaction, $O_{ad} + CO \rightarrow CO_2$, takes place very fast. In contrast, the reaction between CO_{ad} and O_2 gas molecules takes place very slowly at room temperature. The CO adsorbed on Pd surface reacts only with chemisorbed oxygen and the adsorption of oxygen is inhibited by the CO already adsorbed on the surface. The reaction between chemisorbed oxygen and chemisorbed carbon monoxide proceeds slowly at room temperature, but at higher temperatures, it reacts via the Langmuir–Hinshelwood mechanism.

One method of studying the reaction at very low pressures is to pump continuously the system and to leak the reactants in very slowly. When a 3:1 mixture of CO and O_2, at a total pressure of 4×10^{-7} Torr, is introduced onto a palladium surface at 90°C, the Pd surface is almost completely covered by CO, and very little CO_2 is produced. When the CO gas supply is stopped, adsorbed CO starts to desorb and a part of the surface becomes covered by oxygen, which leads to the slow reaction between $CO_{ad} + O_{ad} \rightarrow CO_2$ as given in Fig. 20.

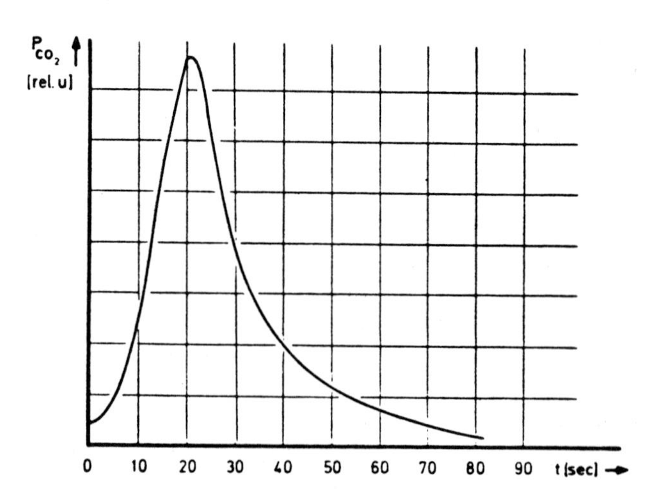

Fig. 20. CO_2 formation through the Langmuir–Hinshelwood mechanism on Pd. Height of the CO_2 peaks as a function of time.[32]

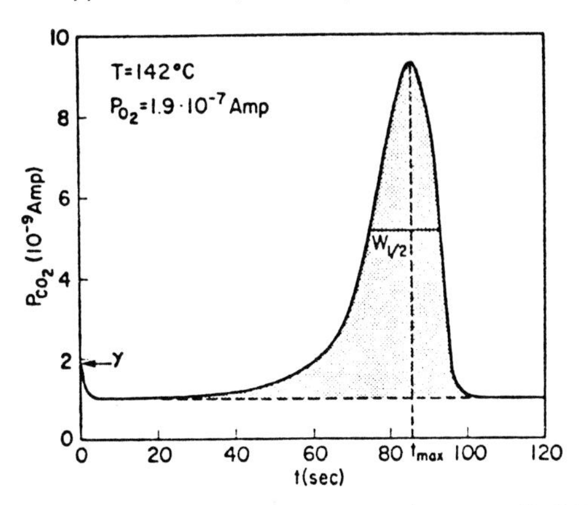

Fig. 21. Plot of the partial pressure of CO_2 versus time over Pt(110), sample in O_2 atmosphere at $T = 142°C$. This curve is indicative of a Langmuir–Hinshelwood mechanism.[31]

In the case of Pt, Bonzel and Ku studied the oxidation reaction on a clean Pt(110) surface by LEED and Auger electron spectroscopic techniques.[31] When the reaction was in steady-state, the partial pressure of CO was suddenly decreased and subsequent CO_2 formation was followed as a function of time as shown in Fig. 21. At these temperatures (142°C) the coverage of CO is appreciable at the moment of the perturbation. Figure 21 indicates that the rate of CO_2 formation drops from its steady-state value γ and then increases gradually, going through a sharp maximum before dropping quickly to a new

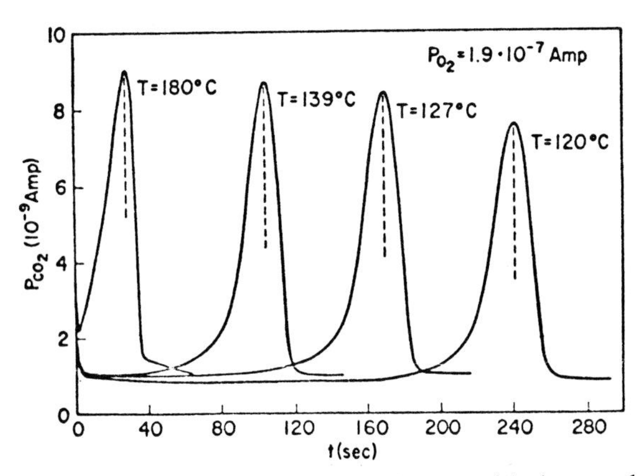

Fig. 22. The occurrence of a maximum in P_{CO_2} changes with the sample temperature; shown are a set of Langmuir–Hinshelwood reaction curves at four different temperatures.[31]

steady-state value dictated by the background pressure of CO. The area under the curve is proportional to the amount of adsorbed CO at $t = 0$. Figure 22 shows the curves obtained at different temperatures, which indicate that the maximum rate of CO_2 production occurs at shorter and shorter times as the temperature is increased.

The behaviour of the reaction over Pd as shown in Fig. 20 and over Pt as in Figs 21 or 22 is very similar in each case and may be interpreted in terms of a Langmuir–Hinshelwood mechanism as follows. At the moment of perturbation, $t = 0$, the partial pressure of CO decreases rapidly to a very low value, and the subsequent formation of CO_2 takes place between the adsorbed CO and adsorbed oxygen, its rate being approximately expressed as:

$$d(CO_2)/dt = k(O_{ad})(CO_{ad})$$

At the moment of perturbation the surface is mainly covered by CO_{ad} which is slowly replaced by O_{ad} to give a maximum value of $(O_{ad})(CO_{ad})$. This is an example of a Langmuir–Hinshelwood mechanism, reaction taking place between competitively chemisorbed reactants.

The rate of formation of CO_2 on Pd in the steady-state of the reaction was examined at various pressures and temperatures. Figure 23 gives the results on various crystal faces and a wire as a function of temperature at constant CO and O_2 pressures. The figure demonstrates that the single crystal faces and the polycrystalline wire exhibit identical catalytic activities. Consequent-

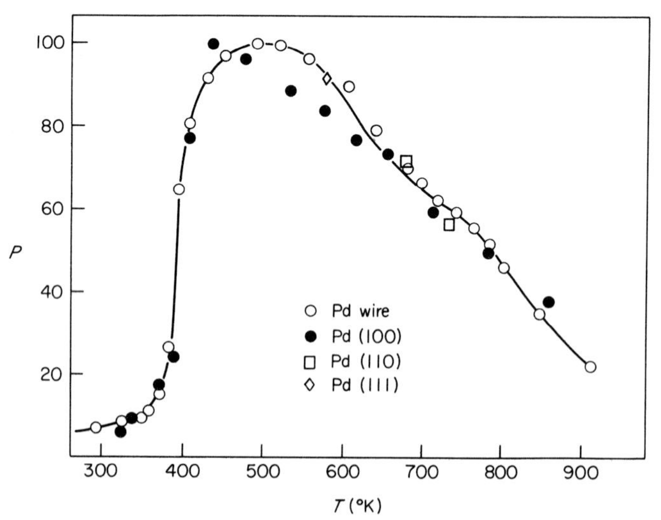

Fig. 23. Steady-state rate of CO_2 formation r as a function of temperature.[32]

ly, it is of great interest to note that no crystal-plane specificity is shown either for the catalytic activity for oxidation or the adsorption.

The amount of CO adsorption under the steady-state of the oxidation reaction was also estimated by means of a laser-induced thermal desorption technique, as shown in Fig. 24.[37] It revealed that CO_2 formation starts when the carbon monoxide coverage becomes smaller than 2/3 of saturation, and while the chemisorption of oxygen is inhibited by adsorbed CO the reaction

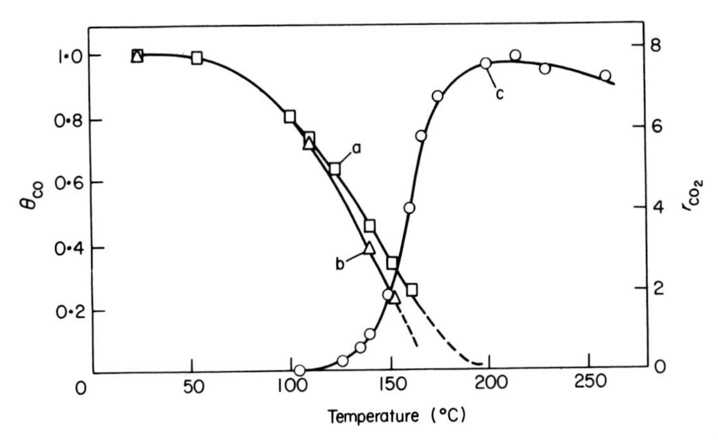

Fig. 24. The rate of CO_2 formation (r_{CO_2}) and CO coverage (θ_{CO}) as a function of temperature.[37]
Curve a: θ_{CO} for $P_{CO} = 8 \times 10^{-9}$ Torr in the absence of O_2.
Curve b: θ_{CO} during the course of the reaction in the stationary state ($P_{O_2} = P_{CO} = 8 \times 10^{-9}$ Torr).
Curve c: r_{CO_2} (relative unit) at $P_{O_2} = P_{CO} = 8 \times 10^{-9}$ Torr.

proceeds predominantly via an Eley–Rideal mechanism, $O_{ad} + CO \rightarrow CO_2$. The formation of CO_2 also takes place via a Langmuir–Hinshelwood mechanism, $O_{ad} + CO_{ad} \rightarrow CO_2$, and the kinetics of this reaction was studied separately by means of a non-stationary pressure jump method.

Accordingly, the following kinetic equations may be written:

$$CO + * \underset{k_{-1}}{\overset{k_{+1}}{\rightleftarrows}} CO_{ad} \tag{4-7-1}$$

$$O_2 + 2* \underset{k_{-2}}{\overset{k_{+2}}{\rightleftarrows}} 2O_{ad} \tag{4-7-2}$$

$$O_{ad} + CO \overset{k_{+3}}{\rightarrow} CO_2 \tag{4-7-3}$$

$$O_{ad} + CO_{ad} \xrightarrow{k_{+4}} CO_2 \qquad (4\text{-}7\text{-}4)$$

where $*$ denotes a free adsorption site for which oxygen and carbon monoxide compete. The total rate of CO_2 formation is then given by

$$R = (O_{ad})[k_{+3}P_{CO} + k_{+4}(CO_{ad})] \qquad (4\text{-}7\text{-}5)$$

At higher CO pressures the first term in the brackets will greatly exceed the second term. If the Langmuir–Hinshelwood mechanism is negligible,

$$R \approx k_{+3}(O_{ad})P_{CO} \qquad (4\text{-}7\text{-}6)$$

At temperatures below 200°C the following approximate relation may be derived for the rate of the steady-state reaction:

$$R \approx \frac{2k_{+2}k_{-1}}{k_{+1}} \frac{P_{O_2}}{P_{CO}} \qquad (4\text{-}7\text{-}7)$$

Under these circumstances, R should be proportional to P_{O_2} and inversely proportional to P_{CO}. Such a kinetic equation has been experimentally obtained by several investigators with "real" catalysts.

From the kinetic expression given by equation (4-7-7) it is obvious that the ratio of the partial pressures of the reactants is always the important factor, rather than their absolute values. This implies that the results obtained at very low pressures may be directly related to studies at higher total pressures provided the ratio of the partial pressures remains similar.

The results demonstrate no significant differences either in the catalytic activities or the adsorption (bridge binding) for different surface orientations. The formation of the chemisorption bond at the surface seems to destroy the coupling of the metal orbitals between the neighbouring metal atoms, which corresponds to the model of "surface molecule" formation. Consequently, the crystallographic orientation becomes of minor importance.

White and his co-workers also studied the oxidation of carbon monoxide over palladium. They confirmed that at temperatures below 200°C the reaction rate is retarded by CO, and above the temperature the reaction proceeds via Eley–Rideal type reaction between chemisorbed oxygen atoms and gas phase CO, the rate increasing rapidly by desorption of CO which allows the chemisorption of O_2. At still higher temperatures the rate falls off due to the decreasing amounts of chemisorbed oxygen atoms available at the surface. They also studied the kinetics of the reaction more extensively and demonstrated that the order of the reaction with respect to O_2 and CO depends upon the partial pressure ratio rather than the total pressure, and markedly upon the reaction temperature in a complex manner. The amount of adsorbed oxygen was measured under working conditions and the kinetic behaviour was explained in terms of a change in the rate-determining step, depending upon the amount of adsorbed oxygen.

In this manner the oxidation of carbon monoxide on clean and well-defined palladium and platinum surfaces has been investigated and its reaction mechanism was elucidated by means of non-stationary techniques and adsorption measurements during the course of the reaction.

The electron spectroscopy, which has been extensively developed in recent years, will certainly be applied to various systems, not only as a tool for investigating chemisorption, an increasingly popular use in recent years, but also as a tool for the dynamic elucidation of the mechanisms of heterogeneous catalysis.

References

1. R. P. Eischens and W. A. Pliskin, *Adv. in Catalysis*, **10,** 1, (1958); L. H. Little, "Infrared Spectra of Adsorbed Species", Academic Press, New York (1966); M. L. Hair, "Infrared Spectroscopy in Surface Chemistry", Marcel Dekker, N.Y. (1967).
2. R. P. Eischens, W. A. Pliskin and M. J. D. Low, *J. Catalysis*, **1,** 180 (1962).
3. C. C. Chang and R. J. Kokes, *J. Am. Chem. Soc.* **93,** 7107 (1971).
4. E. U. Condon, *Phys. Rev.* **41,** 759 (1932); M. F. Crawford and I. R. Dagg, *ibid.* **91,** 1569 (1953).
5. N. Sheppard and D. J. C. Yates, *Proc. Roy. Soc.* **238A,** 69 (1957).
6. S. Naito, H. Shimizu, E. Hagiwara, T. Onishi and K. Tamaru, *Trans. Faraday Soc.* **67,** 1519 (1971); *Bull. Chem. Soc. Japan*, **43,** 974 (1970).
7. R. L. Burwell, Jr., G. L. Haller, K. C. Taylor and J. F. Read, *Adv. in Catalysis*, **20,** 1 (1969).
8. A. L. Dent and R. J. Kokes, *J. Am. Chem. Soc.* **92,** 1092, 6709, 6718 (1970).
9. W. M. H. Sachtler, *Rec. Trav. Chim.* **2,** 243 (1963); C. C. McCain, G. Gough and G. W. Godin, *Nature*, **198,** 989 (1963); H. H. Voge and C. R. Adams, *Adv. in Catalysis*, **17,** 154 (1967).
10. A. L. Dent and R. J. Kokes, *J. Phys. Chem.* **74,** 3653 (1970); **73,** 3772, 3781 (1969).
11. S. Naito, T. Kondo, M. Ichikawa and K. Tamaru, *J. Phys. Chem.* **76,** 2184 (1972).
12. T. Kondo, S. Saito and K. Tamaru, *J. Am. Chem. Soc.* **96,** 6857 (1974).
13. T. Kondo, M. Ichikawa, S. Saito and K. Tamaru, *J. Phys. Chem.* **77,** 299 (1973).
14. Y. Sakurai, T. Onishi and K. Tamaru, *Bull. Chem. Soc., Japan*, **45,** 980 (1972); *Trans. Faraday Soc.* **67,** 3094 (1971).
15. J. F. Woodman and H. S. Taylor, *J. Am. Chem. Soc.* **62,** 1393 (1940).
16. A. L. Dent and R. J. Kokes, *Adv. in Catalysis*, **22,** 1 (1972); R. J. Kokes, *Accnts Chem. Res.* **6,** 226 (1973).
17. S. Naito, Y. Sakurai, H. Shimizu, T. Onishi and K. Tamaru, *Bull. Chem. Soc., Japan*, **43,** 2274 (1970); *Trans. Faraday Soc.* **67,** 1529 (1971).

18. J. M. Trillo, G. Munuera and J. M. Criado, *Catalysis Rev.* **7,** 51 (1973); J. Fahrenfort, L. L. Van Reijen and W. M. H. Sachtler, "The Mechanism of Heterogeneous Catalysis" (J. H. de Boer *et al.*, eds), Elsevier, Amsterdam (1960).
19. Y. Noto, K. Fukuda, T. Onishi and K. Tamaru, *Trans. Faraday Soc.* **63,** 2300 (1967); *Bull. Chem. Soc., Japan,* **40,** 2459 (1967); K. Fukuda, Y. Noto, Y. Onishi and K. Tamaru, *ibid.* **63,** 3072 (1967).
20. Y. Noto, K. Fukuda, T. Onishi and K. Tamaru, *Trans. Faraday Soc.* **63,** 3081 (1967); *Bull. Chem. Soc. Japan,* **40,** 2722 (1967).
21. K. Fukuda, S. Nagashima, Y. Noto, T. Onishi and K. Tamaru, *Trans. Faraday Soc.* **64,** 522 (1968); K. Fukuda, T. Onishi and K. Tamaru, *Bull. Chem. Soc., Japan,* **42,** 1192 (1969).
22. J. J. F. Scholten, P. Mars, P. G. Menon and R. van Hardeveld, "Proc. 3rd Int. Congr. Catalysis", Amsterdam, p. 881 (1964).
23. A. Ueno, T. Onishi and K. Tamaru, *Trans. Faraday Soc.* **66,** 756 (1970); *Bull. Chem. Soc., Japan,* **42,** 3040 (1969).
24. P. W. Darby and C. Kemball, *Trans. Faraday Soc.* **53,** 832 (1957); A. J. Dandy, *J. Chem. Soc.* **1963,** 5956; T. Saida and A. Ozaki, *Bull. Chem. Soc., Japan,* **37,** 1817 (1964); R. O. Kagel and R. G. Greenler, *J. Chem. Phys.* **49,** 1638 (1968); Y. Soma, T. Onishi and K. Tamaru, *Trans. Faraday Soc.* **65,** 2215 (1969).
25. A. Ueno, T. Onishi and K. Tamaru, *Trans. Faraday Soc.* **67,** 3585 (1971).
26. A. C. Herd, T. Onishi and K. Tamaru, *Bull. Chem. Soc. Japan,* **47,** 575 (1974).
27. I. Langmuir, *Trans. Faraday Soc.* **17,** 621 (1922).
28. H. Heyne and F. C. Tompkins, *Proc. Roy. Soc.* **292A,** 460 (1966); M. Akhtar and F. C. Tompkins, *Trans. Faraday Soc.* **67,** 2461 (1971).
29. B. J. Wood, N. Endow and H. Wise, *J. Catalysis,* **18,** 70 (1970).
30. G.-M. Schwab and K. Gossner, *Z. physik. Chem., N.F.* **16,** 39 (1958).
31. H. P. Bonzel and R. Ku, *J. Vac. Sci. Tech.* **9,** 663 (1972); *Surf. Sci.* **33,** 91 (1972).
32. G. Ertl and J. Koch, "5th Int. Congr. Catalysis", **67,** 969 (1974).
33. R. L. Park and H. H. Madden, *Surf. Sci.* **11,** 188 (1968).
34. G. Ertl and K. Koch, *Z. physik. Chem., N.F.* **69,** 323 (1970).
35. R. F. Baddour, M. Modell and U. K. Heusser, *J. Phys. Chem.* **72,** 3621 (1968).
36. B. Lang, R. W. Joyner and G. A. Somorjai, *Surf. Sci.* **30,** 454 (1972); G. A. Somorjai, *Catalysis Rev.* **7,** 87 (1973).
37. G. Ertl and M. Neumann, *Z. physik, Chem., N.F.,* **90,** 127 (1974): G. Ertl, *Angew. Chem.* (*Internl. ed.*) **15,** 391 (1976).
38. J. S. Close and J. M. White, *J. Catalysis,* **36,** 185 (1975); T. Matsushima and J. M. White, *J. Catalysis,* **39,** 265 (1975); T. Matsushima, D. B. Almy, D. C. Foyt, J. S. Close and J. M. White, *J. Catalysis,* **39,** 277 (1975); T. Matsushima, C. J. Mussett and J. M. White, *J. Catalysis,* **41,** 397 (1976).

Subject Index